First printing, June 2014

To my spouse, Christine:
Tu ventus sub alis meis es.

And to my first statistics professor, Randy Spoeri:
You introduced me to the subject and made it fun.

Contents

Descriptive Statistics & Data Science

1   Introduction 3

1.1   Data Science & Statistics   **4**

1.1.1   Why Study Probability and Statistics? 5

1.2   More On Statistics   **6**

1.2.1   Populations and Samples 7

1.2.2   Descriptive versus Inferential Statistics 8

1.2.3   How to Study Statistics 8

1.3   An Introduction to R   **10**

1.3.1   Getting Started 10

1.3.2   Some Important R Paradigms 15

1.3.3   Useful References and Getting Help 18

1.3.4   Extending R: Installing Packages 20

1.3.5   A Few Final Notes 21

1.4   Problems   **24**

2   Descriptive Statistics 25

2.1   Introduction   **25**

2.1.1   Types of Data 26

2.1.2   Example Data: U.S. Domestic Flights from 1987 to 2008 27

2.2   Cross-sectional Data   **30**

2.2.1   Measures of Location 30

2.2.2	Measures of Variation	35
2.2.3	Measures of How Two Variables Co-vary	38
2.2.4	Other Summary Statistics	41
2.3	**Longitudinal Data**	**45**
2.3.1	Statistics for Repeating Cross-sections of Data	45
2.3.2	Statistics for Moving Windows of Data	46
2.4	**Tabular Summaries of Data**	**50**
2.5	**Problems**	**53**
3	**Data Visualization**	**55**
3.1	**Introduction**	**55**
3.2	**Traditional Statistical Graphics**	**56**
3.2.1	Bar Charts	56
3.2.2	Histograms	60
3.2.3	Lattice (or Trellis) Plots	65
3.2.4	Box Plots	67
3.2.5	Scatterplots	71
3.3	**Graphics for Longitudinal Data**	**75**
3.3.1	Time Series Plots	75
3.3.2	Repeated Cross-sectional Plots	76
3.3.3	Autocorrelation Plots	78
3.4	**Other Types of Data Visualization**	**79**
3.4.1	Text Visualization	80
3.4.2	Survey Data Visualization	80
3.4.3	Geo-spatial Visualization	82
3.4.4	Network Visualization	83
3.5	**Problems**	**86**

Probability

4	**Basic Probability**	**91**
4.1	**Introduction**	**91**
4.2	**Events and Sample Spaces**	**92**
4.2.1	Probability Axioms	93
4.2.2	Union of Events	93
4.2.3	Intersection of Independent Events	96
4.2.4	Complementary Events	98
4.2.5	Conditional Probability	99

4.3 **Calculating Probabilities**   **101**

4.3.1   Sample Point Method 102

4.3.2   Counting Sample Points 104

4.3.3   Combining Events 110

4.4 **Bringing It All Together**   **111**

4.4.1   Law of Total Probability 111

4.4.2   Bayes' Theorem 114

4.5 **Problems**   **117**

5   Random Variables 121

5.1 **Introduction**   **121**

5.2 **Discrete**   **122**

5.2.1   Probability Mass Function 122

5.2.2   Cumulative Distribution Function 126

5.2.3   Expected Value 129

5.2.4   Variance and Standard Deviation 131

5.3 **Continuous**   **134**

5.3.1   Probability Density Function 134

5.3.2   Cumulative Distribution Function 138

5.3.3   Expected Value 142

5.3.4   Variance and Standard Deviation 142

5.4 **Expected Value Properties**   **143**

5.5 **Variance Properties**   **145**

5.6 **Problems**   **145**

6   Discrete Distributions 147

6.1 **Introduction**   **147**

6.2 **Binomial Distribution**   **147**

6.3 **Geometric Distribution**   **155**

6.4 **Negative Binomial**   **159**

6.5 **Hypergeometric Distribution**   **160**

6.6 **Poisson**   **160**

6.7 **Problems**   **163**

7   Continuous Distributions 165

7.1 **Introduction**   **165**

7.2 **The Normal Distribution**   **165**

7.2.1   Standardizing 167

7.2.2	Bivariate and Multivariate Normal Distributions	167
7.3	**Other Continuous Distributions**	**168**
7.3.1	Exponential	168
7.3.2	Chi-squared (χ^2)	168
7.3.3	F Distribution	168
7.3.4	t Distribution	170
7.4	**Tchebysheff's Theorem**	**171**
7.5	**Problems**	**171**
8	**Sampling Distributions and the Central Limit Theorem**	**173**
8.1	**Introduction**	**173**
8.2	**Sampling Distributions**	**173**
8.3	**The Central Limit Theorem**	**174**
8.4	**Proof of the Central Limit Theorem**	**174**
8.5	**Normal Approximation to the Binomial**	**174**
8.6	**Problems**	**174**

Estimation & Inference

9	**Data Collection and Sampling**	**179**
9.1	**Introduction**	**179**
9.2	**Why Sample?**	**179**
9.3	**Simple Random Sampling**	**179**
9.4	**Stratified Sampling**	**179**
9.5	**Cluster Sampling**	**179**
9.6	**Other Types of Sampling**	**179**
9.7	**Problems**	**179**
10	**Point Estimation**	**181**
10.1	**Introduction**	**181**
10.2	**Bias and Mean Square Error**	**181**
10.3	**Finding Point Estimators**	**181**
10.4	**Assessing Point Estimators**	**181**
10.5	**Some Unbiased Point Estimators**	**181**
10.6	**Problems**	**181**

11 Confidence Intervals 183

11.1 Introduction   183

11.2 Basic Properties   183

11.3 Large Sample Confidence Intervals   183

11.4 Small Sample Confidence Intervals   183

11.5 Bootstrap Confidence Intervals   183

11.6 Confidence Intervals for the Variance and Standard Deviation   183

11.7 Problems   183

Hypothesis Testing

12 Hypothesis Tests for One Sample 187

12.1 Introduction   187

12.2 Elements of a Statistical Test   187

12.3 Testing the Mean   187

12.4 Testing the Proportion   187

12.5 Testing the Variance   187

12.6 P-values   187

12.7 Bootstrapping Hypothesis Tests   187

12.8 Problems   187

13 Hypothesis Tests for Two Samples 189

13.1 Introduction   189

13.2 Testing Independent Samples   189

13.3 Testing Paired Samples   189

13.4 Confidence Intervals for Two Samples   189

13.5 Problems   189

14 Hypothesis Tests for Discrete Data 191

14.1 Introduction   191

14.2 Goodness-of-Fit Tests   191

14.3 Chi-square Tests   191

14.4 Problems   191

15 Analysis of Variance 193

15.1 Introduction   193

15.2 Single-factor ANOVA   193

15.3 Multiple Comparisons in ANOVA   193

15.4 Two-factor ANOVA   193

15.5 Multi-factor ANOVA   193

15.6 Introduction to Design of Experiments   193

15.7 Problems   193

16 Nonparametric Hypothesis Testing 195

16.1 Introduction   195

16.2 Wilcoxon Signed-Rank Test   195

16.3 Wilcoxon Rank-Sum Test   195

16.4 Mann-Whitney U Test   195

16.5 Kruskall-Wallis Test   195

16.6 Friedman Test   195

16.7 The Runs Test   195

16.8 Rank Correlation Coefficient   195

16.9 Problems   195

Modeling

17 Simple Linear Regression 199

17.1 Introduction   199

17.2 The Regression Model   199

17.3 Method of Least Squares   199

17.4 Inference on the Intercept (β_0) and Slope (β_1) Parameters   199

17.5 Prediction Intervals for Future Observations   199

17.6 Correlation   199

17.7 Problems   199

18 Nonlinear and Multiple Regression 201

18.1 Introduction   201

18.2 Multiple Regression Model   201

18.3	Model Checking and Validation	201
18.4	Polynomial and Nonlinear Regression	201
18.5	Issues in Multiple Regression	201
18.6	Problems	201

19	Time Series Models	203
19.1	Introduction	203
19.2	Smoothing Models	203
19.3	Regression-based Models	203
19.4	ARMA and ARIMA Models	203
19.5	Problems	203

A	More About R	205
A.1	Introduction	205
A.2	Reading Data into R	205
A.3	Managing Your Workspace	205
A.4	Writing Scripts	205
A.5	Writing Functions	205

| | Index | 211 |

Preface

Data science is a new field that has arisen to exploit the proliferation of data in the modern world. Statistics dates back to the mid-18th century, where the field began as the systematic collection of population and economic data by nations. The modern practice of statistics – which includes the collection, summarization, and analysis of data – dates to the early 20th century. Today statistical methods are widely used by governments, businesses and other organizations, as well as by all scientific disciplines.

In many ways, data science is really just the next step in the development of the discipline of statistics. Traditional statistics arose in an era in which data was hard (and thus expensive) to collect and so traditional statistical methods were created to extract the most information possible from data. Today, with the proliferation of computers, sensors, and the internet, some types of data are cheap and plentiful. This does not mean that the traditional statistical methods are now obsolete – they still have much to contribute to today's data-rich environment – but they do require appropriate application.

This book is for data scientists who want to improve how they work with, analyze, and extract information from data. It is intended for current and future data scientists alike, and for anyone interested in deriving information from data. It requires some mathematical sophistication on the part of the reader, as well as comfort using computers and statistical software, but for data scientists these prerequisites should be a given.

It has been said that a data scientist must have a better grasp of statistics than the average computer scientist and a better grasp of programming than the average statistician. This book will give the data scientist a basic grasp of statistics. It should be the first step on a longer journey to master statistical methods, a field which is both broad and deep. As the Chinese philosopher Laozi said, "A journey of a thousand miles begins with a single step."

Statistics is often viewed by students with either trepidation or distaste. This is unfortunate because applied statistics, which is nothing more than using data to find the answer a question or solve a scientific mystery, can be both interesting and a lot of fun. As Francis Galton said in 1889,

> *Some people hate the very name of statistics, but I find them full of beauty and interest. Whenever they are not brutalized, but delicately handled by the higher methods, and are warily interpreted, their power of dealing with complicated phenomena is extraordinary. They are the only tools by which an opening can be cut through the formidable thicket of difficulties that bars the path of those who pursue the Science of Man.*

The focus of this book is how to appropriately apply statistical methods, both simple and sophisticated, to 21st century data and problems. Because differences in assumptions and methods can produce divergent results, data scientists must be statistically savvy enough to understand, interpret, and reconcile ensuing differences in conclusions. Only then can they separate fact from opinion, objective analysis from biased agenda, good data science from bad.

Monterey, California
June 2014

R.D. Fricker, Jr.
Professor

Acknowledgements

This book was typeset in LaTeX using a modified version of The Legrand Orange Book template originally created by Mathias Legrand and modified by Vel and the author.

Descriptive Statistics & Data Science

1 — Introduction

In the 1990s and before, most of the world's information was stored on paper and other analog media such as film. However, with the proliferation of personal computers and the internet, by 2000 one quarter of the world's information was stored digitally. Since that time, the amount of digital data has exploded, roughly doubling every couple of years, so that now *more than 98 percent of all stored information is digital*.

Much of this digital data is the result of the *datafication* of the world. Datafication is both the digitization of existing analog media and, more significantly, the collection of digital data on people, processes, and other things in ways that until recently were not possible. For example, the rise of social media has resulted in the generation of massive amounts of digital data by and about individuals throughout the world. More generally, the ubiquitous proliferation of smart sensors and ever-cheaper storage is driving the availability of data from all types of societal, commercial, and government processes and systems. The result is an exponentially increasing amount of data being collected and stored, much of which is in need of analysis so that useful information can be extracted from the data.

What does some of this data look like? In October of 2013, *Wired* magazine said,

> *Every day, we collectively produce millions of books' worth of writing. Globally, we send 154.6 billion emails, more than 400 million tweets, and over 1 million blog posts and around 2 million blog comments on WordPress. On Facebook, we post about 16 billion words. Altogether, we compose some 52 trillion words every day on e-mail and social media—the equivalent of 520 million books. (The entire US Library of Congress, by comparison, holds around 23 million books.)*

This flood of digital data contains valuable information that can be used to inform all types of decisions. Indeed, cutting-edge commercial organizations now focus their operations around the analysis and exploitation of knowledge gleaned from data. This is what data science is all about: Turning data into useful information.

1.1 Data Science & Statistics

Both data science and statistics are concerned with of the extraction of useful information from data. Let's start by defining the two fields.

> **Definition 1.1.1 — Statistics.** Statistics is the science of learning from data, including collecting, organizing, analyzing, interpreting and presenting data, often with a particular focus on measuring, controlling, and communicating the uncertainty inherent in the data, associated analyses, and final results or conclusions.

Statistics traces its roots back to 2 A.D. when the Han dynasty conducted a census of the Chinese population, where it counted 57.7 million people in 12.4 million households. As this early application illustrates, statistics is about both the collection of data as well as its analysis. Furthermore, as Definition 1.1.1 makes clear, statistics is also focused on determining the uncertainty in data. This is an important consideration when only a sample of data is observed because the results will be subject to sampling error, and classical statistics is generally predicated on the idea that it's not possible to observe all the data, either because it's too expensive or it's impossible to collect it all. We'll explore this concept more in Section 1.2, but briefly the idea is to use the data not only to summarize what was observed but also to quantify what can be said about *all* of the data, both observed *and* unobserved.

Statistics can also be split into two broad sub-fields: theoretical statistics and applied statistics. Theoretical statistics is concerned with the creation and development of methods and techniques to summarize and analyze data, including clearly defining how and when to use the methods and their associated pros and cons. Applied statistics, on the other hand, is the application of the methods to data and the process of conducting rigorous and principled analyses. In the statistical profession, the division between applied and theoretical statisticians is very fuzzy, with most statisticians doing both, though perhaps with an emphasis on one or the other.

When developing methods, statisticians seek to understand how the methods perform on particular types of data, including how efficiently they extract information from the data and how well they characterize the uncertainty inherent in a sample of data. That is, just as automobile designers seek to understand the performance characteristics of a new car, statisticians seek to understand how their methods perform. In particular, as Lindsay et al. say, "A distinguishing feature of the statistics profession, and the methodology it develops, is the focus on a set of cautious principles for drawing scientific conclusions from data."[1]

> **Definition 1.1.2 — Data Science.** Data science the study of how to extract useful information from data using quantitative methods and theories from many fields, including statistics, operations research, computer science, and various engineering disciplines. Data science often focuses on large data sets not originally designed or collected to address the question of interest.

In many ways data science is a modern extension of statistics and, to the extent they use statistical methods, data scientists can be characterized as applied statisticians. Indeed, while some trace the inception of data science to the 1960s, where it was then focused on data processing, modern data science originated with a lecture given by Professor C.F. Jeff Wu in 1998 entitled "Statistics = Data Science?" In that lecture, Professor Wu characterized statistical work as data modeling, analysis,

---

[1]Lindsay, B.G., Kettenring, J., Siegmund, D.O. (2004). A Report on the Future of Statistics, *Statistical Science*, **19**(3):387–413.

and decision making where he proposed that statistics be renamed data science and statisticians be called data scientists.

However, since the late 1990s, particularly with the explosion of massive, heterogeneous, and often unstructured data sets, the term data science has expanded to include the ability to collect, manage, and analyze such data. Today, data scientists are expected to be adept in both statistics and computer science, particularly as applied to extracting and manipulating large data sets, as well as have a solid working knowledge of the field in which they are trying to answer questions. In particular, data scientists must be able to:

- find, manage, and interpret large and complex data sets,
- analyze the data, including building mathematical models,
- present and communicate results.

As a result, as Definition 1.1.2 makes clear, today's data scientists come from a variety of fields and academic backgrounds and they collect and analyze data using a variety of methods. Thus, data science now extends beyond the realm of traditional statistics that was generally focused on collecting and analyzing smaller and typically very structured numerically-based data sets. Yet, coming full circle, those who collected and collated the census data back in 2 A.D. for the Han dynasty – where collecting information on 57.7 million people was undoubtedly a huge undertaking that resulted in a massive amount of data for that era – could have been called data scientists!

1.1.1 Why Study Probability and Statistics?

Why should a data scientist care about learning statistics? It has often been said that a good data scientist knows more statistics than the typical computer scientist, and he or she knows more computer science than the typical statistician. That is, a data scientist must be both adept at computer science in order to deal with the necessary data and the data scientist must also know how to rigorously analyze the data to reach the appropriate conclusions.

Successful data science requires skills that fall into three broad areas:

- Statistical methods, including knowledge of probability, sampling, estimation, hypothesis testing, and multivariate modeling and analysis.
- Statistical concepts, including the foundations of inference, accuracy and precision of estimates, particularly the sources and types of bias, and understanding the difference between correlation and causation, .
- Computer science methods, including the handing and storage of large data sets, databases, computational algorithms, and distributed, parallel and fault-tolerant computing.

The first two bullets are precisely within the purview of statistics, where there are four ways statistics can be used to address data science questions. Statistical methods can be used to:

- *summarize* a large body of data in useful ways;
- *discover* new relationships in data;
- *test* theories and explanations with empirical evidence; and,
- *confirm* that relationships in the data are unlikely to be the result of chance.

Sophisticated statistical techniques are increasingly necessary for data science's complicated problems. And even the most basic problems require a modicum of statistical literacy, if only to ensure that the resulting recommendations have been arrived at in a reasonable and unbiased manner.

Furthermore, for data scientists working with massive amounts of data, it is frequently *not* sufficient to answer a complex problem with a pie chart – indeed, such an approach can often be misleading and result in incorrect conclusions. Data scientists must also be statistically savvy enough to understand and interpret, as well as appropriately reconcile or adjudicate, differing analyses and results.

This book will present statistical methods to do all of these things. It is written for those who seek to improve the practice of data science. The goal of this book is not to turn data scientists into a statistician. Rather, it is to give data scientists an introduction to the field and to provide data scientists with the skill to:

- *apply* basic statistical methods to data science problems;
- *understand* more advanced statistical techniques and how they are properly applied;
- *judge* good statistics and statistical practice from bad; and,
- *know* when to call on statistical experts.

1.2 More On Statistics

While most of the colloquial and popular media references to statistics concern the collection and summarization of a set of numbers (e.g., baseball statistics or stock market returns), real statistics is about much more than that. In fact, if the field of statistics was only concerned with describing data, which is referred to as *descriptive statistics*, this book would conclude with Chapter 3.

Statistics is most fundamentally about methods for describing uncertainty. For example, uncertainty may arise if a data science question is about a particular *population* but data are only available on a subset of the population – a *sample*. Hence, there is uncertainty about how closely the results from the sample correspond to the results for the population. Similarly, uncertainty may arise if the data science question involves forecasting the future which, of course, can only be answered using data from the past and present.

For example, what if we wanted to know the average starting salary for a person obtaining a master's degree in data science in the United States? One way to find out would entail getting the salary information for every new data scientist in the U.S. and then calculating the average. The left side of Figure 1.1 illustrates this idea. However, obtaining the starting salary data for every single data scientist in the United States is probably impossible to do. Alternatively, we could collect the starting salary information from a *sample* of new data scientists with master's and use it to *estimate* the average salary of the entire population. The right side of Figure 1.1 illustrates this idea, where the goal is to use the sample results to understand the population, and so it is clearly important to ensure the sample is representative of the population.

Similarly, we might want to forecast what the starting salary of a new data scientist in the United States will be in two years. To answer this question, we might use a dataset containing samples of starting salaries for new data scientists every year for the past ten years and, with it, *predict* the average starting salary two years from now.

In either situation the natural question that arises is "how far off is the estimated or predicted average salary from the actual value?" After all, in Figure 1.1, the sample is *not* the population, so any analysis on sample data is likely to differ from the same analysis done on the complete population's data. Statistical methodology is designed to formally specify the precision and uncertainty inherent in any such estimate or prediction.

Of course, since it is often difficult or impossible to know the true result for the population, it can be easy for unprincipled analysts to fool those not skilled in statistics and hence the infamous

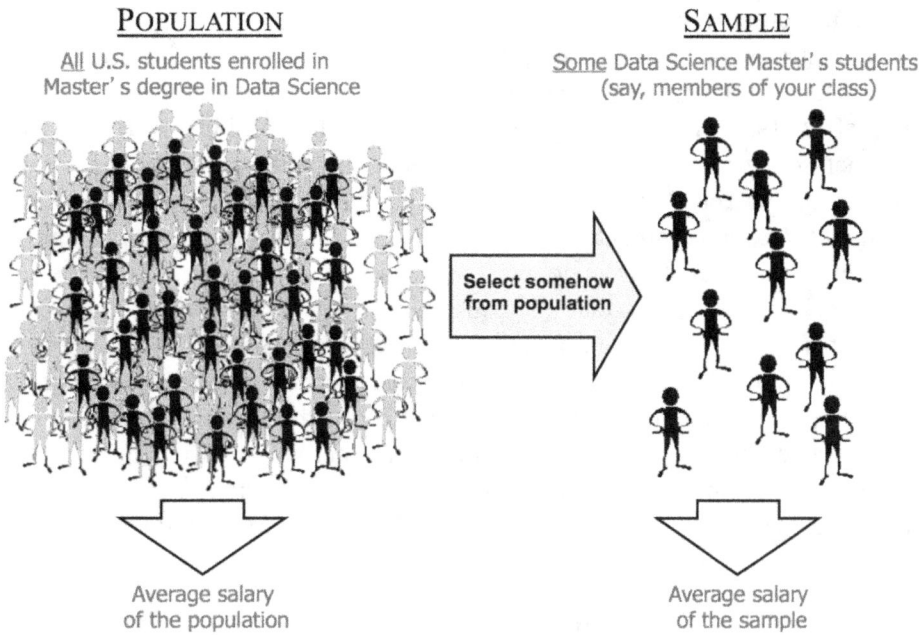

Figure 1.1: Calculating the average starting salary for the entire population of data scientists in the United States versus the average starting salary for a sample of data scientists. Of course, the average calculated from a sample is unlikely to be the same as the average calculated from the population.

saying: "There are three types of lies: lies, damn lies, and statistics." *Good statistics is about defining mathematically rigorous ways to do estimation, hypothesis testing, and modeling combined with principled methods for quantifying how far off an estimate is likely to be from the true answer.* That may sound like a bit of magic, but you will learn how to do it in this book!

1.2.1 Populations and Samples

We used the terms *population* and *sample* to motivate what statistics is all about: quantifying uncertainty. A population is the set of all people or things that meet the criteria of a particular research study or data science question. A sample is a subset of the population upon which the study or analysis will actually be done. A *random sample* is a subset that is not drawn in any systematic way from the population. (We'll learn more about sampling in Chapter 9.)

For example, if we were interested in saying something about the average GRE scores for students studying data science this year, then the population would be all students enrolled in a data science degree program this year. A sample of that population could be the students in your statistics class class. That sample is not likely to be random, however, since it probably systematically excludes certain groups of students (such as students enrolled in data science programs at other schools).

Now, on the other hand, if we are interested in the average height of students in your statistics class, the class is the population. A sample might be all the women in the class. Is that a random sample? If we used the average of the heights of the women in the class to estimate the average height of all students in the class, would we be making a good estimate?

How might we draw a random sample of students in the class? One way would be to have a computer assign each person a random number and then take the n people with the biggest or smallest random numbers. Another way would be to put everyone's name on a 3x5 card (one name to each card), put them in a hat, shuffle them well, and then draw out n of them.

Why sample? Often it is either impossible or financially prohibitive to observe an entire population. In fact, sometimes even with significant resources and extraordinary effort it is difficult to accurately measure an entire population. A good example is the United States Census. Every decade the United States government spends millions of dollars and puts forth significant effort trying to count every individual in the country – a massive undertaking. And, every decade the Census is challenged for failing to accurately count certain segments of society.

As it turns out, via good statistical practice, we can often get as precise answers from a sample as we can from attempting to collect data from the whole population. Furthermore, there are times when taking a sample can be more precise than trying to get the whole population. How can this be? Well, for the same amount of effort or cost one can either get precise data from the sample or imprecise data from the population. The idea is that under certain conditions it is preferable to allow for a moderate increase in *sampling error* in order to achieve a greater reduction in *measurement error*.

1.2.2 Descriptive versus Inferential Statistics

In Chapters 2 and 3, we will learn about descriptive statistics and data visualization. These are ways to numerically and graphically summarize data, whether the data are from a sample or a population. Why is this important? Think about the United States Census with its information on more than 300 million people. If we wanted to understand the economic status of people in the United States we would certainly not want to do so by looking at each and every Census record. Rather, we would use ways to describe the data in a more concise way, either through summary statistics or graphical plots of the data. That is, we would use descriptive statistics to summarize the data.

Most of the rest of this book is about *inferential statistics* (though we will have to spend quite a few chapters developing the probability tools we will need to do statistical inference first). This is the machinery designed for using a *statistic* calculated from a sample of data to say something about the population. As illustrated in Figure 1.2, if it is impossible to obtain the starting salary for every data scientist in the U.S., then we will have to use information from a sample to *infer* what it is for the population. However, inference is also more than using a sample average as an estimate for a population average; it is also about statistical methods to quantify how accurate the sample average is and thereby specify our uncertainty in our knowledge of the population.

1.2.3 How to Study Statistics

Learning statistical material often requires a style of studying that differs from other subjects. With statistics, this is because: (1) the terminology and mathematical details can be overwhelming; (2) the concepts can sometimes be confusing or counterintuitive; and, (3) the ideas often take some quiet, careful thought to understand and internalize. In particular, reading statistics texts is not like reading a case study or a novel. It can be very hard to just start at the beginning and read it straight through. To effectively read this book and learn the material, to do it in two parts:

1. Skim each chapter first, looking for the big ideas that are being presented and the major points. Here you should be asking "what" and "why" questions: "What problem is being solved?" and "Why is it a statistics problem?" and "Why is this particular statistical technique being used to solve this problem?"

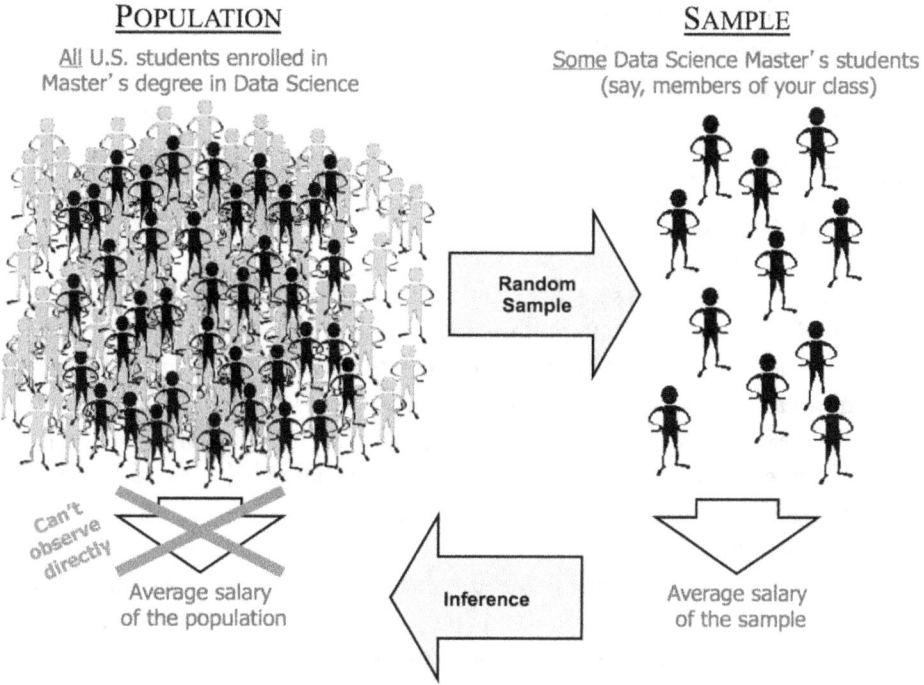

Figure 1.2: Statistical inferential methods are designed to estimate population *parameters* using *statistics* generated from a sample and to quantify the precision of the estimate.

2. Then, once you understand the major idea(s), go back and re-read everything more carefully, asking the "how" and "when" questions: "How does the method work?" and "When do I use the method?" and "How do I actually apply it?"

In addition, one point that cannot be stressed enough is that you should not depend only on interaction with others to learn the material. First, some concepts are difficult, and those doing the explaining may not get them exactly right. Second, other concepts can be a bit slippery. That is, an explanation may make perfect sense from someone else, but you may not really get the concept without careful thought and doing an exercise or two on your own.

If you are using this book as part of a statistics course, consider using the following general study regimen:

1. Skim each reading assignment before the lecture. Don't worry if you don't get all the concepts. It is most important at this point is to understand the main ideas and the concepts that will be presented in class so you will be able to concentrate on the details.
2. As you go through the lecture, stay mentally engaged and ask yourself questions. Use this time to deepen your understanding of the material.
3. After the lecture, carefully do the reading in detail (as described above) and then, if there is an assignment, attempt the assignment. If you have difficulty with completing the assignment, go back and re-read the relevant chapters.
4. Check yourself by doing the exercises and homework assignments.
5. If you continue to have difficulties with the reading or homework, see your professor or your

teaching assistant and attend office hours sessions. If you discuss the homework with your classmates, be sure you can do the assignment on your own, as this will be the true test of how well you understand the material.

1.3 An Introduction to R

R is cutting-edge, free, open-source statistical software. It is used by thousands of people, from academic researchers to wall street analysts. R runs on a wide variety of UNIX platforms, and on the Windows and MacOS operating systems. To download it, go to `http://www.r-project.org/` and follow the downloading and installation instructions.

Unlike much of modern software in which users interact with the software via a graphical user interface (also known as a GUI), R runs from a command line. Thus, in order to interact with it, users must master R's syntax and then type in commands. In this sense, R is similar to writing software code. However, unlike most programming languages that have to be compiled, R is interactive, meaning that commands can and are sequentially entered and this allows the user to observe of the output from each command before executing the next one.

With most forms of statistical analysis this type of interactivity is important because it is often not known what the appropriate next step in an analysis should be until the results of the previous step is observed. This sort of "human-in-the-loop" data analysis is what is most commonly done by statisticians, though note that it is also possible to write and run *R scripts* which act like programs, and these can be very handy for automating tasks for those analytical tasks that do not require human-in-the-loop decision making.

1.3.1 Getting Started

Once you have installed R, boot the program in the usual way for your operating system. In Windows or MacOS, this usually this involves clicking or double clicking on the R icon: ℝ.[2] If R has been successfully installed, you should get a screen that looks something like Figure 1.3 or Figure 1.4. (The exact look varies by operating system and R version.) Note the command prompt, $>$, on the bottom line in Figures 1.3 and 1.4. This is where you enter commands in R to tell the software what to do. After entering a command on the command line, you hit the ENTER or RETURN key to tell R to execute the command.

Exercise 1.1 Install R on your computer and ensure it works.   ■

For example, R can be used as a calculator. Imagine you wanted to add the numbers 52 and 18. At the command prompt you would type 52 + 18 and hit the ENTER key. Below is what you would see in R.

```
> 52 + 18
[1] 70
```

The [1] on the second line indicates the output, where the number 1 inside the brackets means that 70 is first (and in this case the only) number in the output.

R code is composed of *expressions*. Example 1.1 shows a series of arithmetic expressions and their output. Try reproducing these on your computer.

---

[2]In Linux, or if you are working from a Mac terminal window or the DOS line in Windows, just type R.

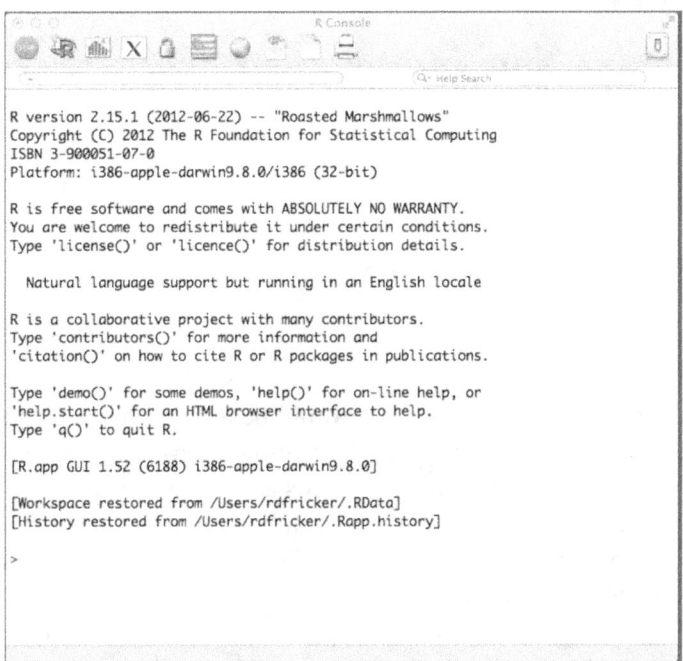

Figure 1.3: Screenshot of R (version 2.15.1) running on a Mac. The R interface is really a very spare GUI.

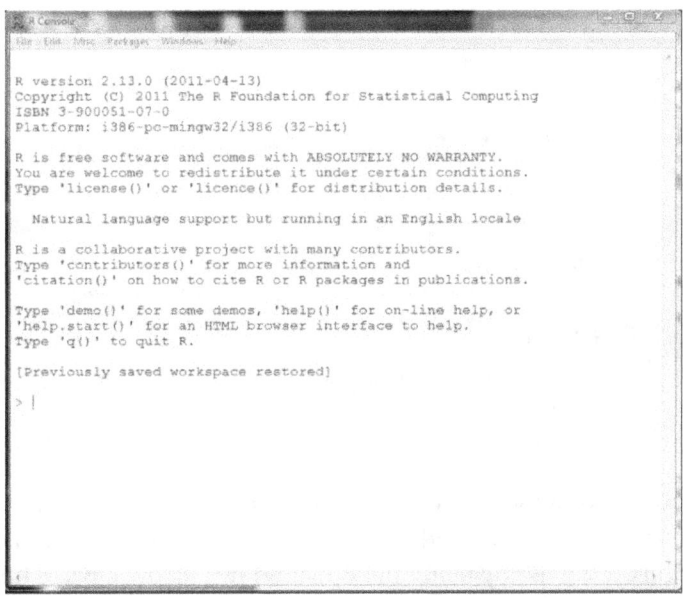

Figure 1.4: Screenshot of R (version 2.13.0) running on a PC. As with the Mac R interface in Figure 1.3, the Windows R interface is also a very spare GUI.

■ **Example 1.1 — Using R as a calculator.**

```
> 1 - 2 * 3
[1] -5
> 4 * (2 + 7)
[1] 36
> 100 / 5
[1] 20
> 3 ∧ 3 +1
[1] 28
> sqrt(10)
[1] 3.162278                                                                          ■
```

Now, if arithmetic was all that R could do, it would be nothing more than a glorified calculator. But, as we will see, R can do much more powerful computations. In reviewing the examples, note that R respects the usual order of arithmetic operations as well as the standard use of parenthesis to specify some desired order. Also note that the last calculation uses the built in function sqrt() to calculate the square root of a number. Here the number we want to take the square root of, 10, is inserted between parentheses as an *argument* to the function.

Of course, these are just a few examples of how to do some basic arithmetic calculations. In addition to the usual arithmetic operators, $+$, $-$, $*$, $/$, and $\wedge$, the equals sign can be used as an *assignment operator* (which we will discuss shortly), and there are logical operators. For example, the operator == (a double equals sign) tests whether two objects are equal, != tests whether they are not equal, and >, <, >=, <= do just what you'd expect.

■ **Example 1.2 — Logical operators.**

```
> 2 * 3 == 6
[1] TRUE
> 1310 >= 26359/20
[1] FALSE
> 24/90 != 8/30
[1] FALSE                                                                             ■
```

> **Exercise 1.2** Solve the following using R.
>   1. $4 + 2.5^4$
>   2. $(52.431 + 27.936) \div (4.569 \times 9.023)$
>   3. $\sqrt{43} \times \left(\frac{51}{87} + \frac{36}{127} \right)$
>
> *Answer*: 43.0625, 1.9494, 5.7028.   ■

A very useful thing to know when working with the R command line is that you can scroll through all the commands you've executed during an R session (and, as we'll soon see, even some you've typed in previous sessions) by using the up and down arrow keys. This can be particularly useful if, for example, you make a typo when typing in a long command. Instead of completely retyping the command to correct the mistake, simply hit the up arrow key to bring the previously entered command back up on the command line and use the sideways arrow keys and the delete key to fix the mistaken part. Once corrected, just hit the enter key (with the cursor anywhere on the command line) to re-execute the (now corrected) command.

Typesetting Conventions

As shown in the previous section, Courier font will be used throughout the text to denote R commands and output. Also, throughout the text, the R examples will be shown with the command prompt but, of course, you do not type that in when running the command. Indeed, doing so will produce an error because R interprets ">" as the greater than symbol.

Note that functions will be displayed with parenthesis after the function name to differentiate them from other R syntax. And, while most functions have at least one required argument and often many optional arguments, in the main body of the text we will generally just show the parentheses without any arguments or with an italicized dummy argument. Of course, in the examples and exercises, the appropriate arguments will be shown. As we will discuss in Section 1.3.3, you can always use the R help pages to get a complete listing of a functions arguments.

Finally, note that R is case-sensitive. For example, the text Some.Text is separate and distinct from some.text.

■ Example 1.3 — R is case sensitive.

```
> "Some.Text" == "some.text"
[1] FALSE
```
■

Similarly, commands and object names are case sensitive. So, when you are following along with the textbook examples, and in your own work, it is important to be conscious of text case.

The Assignment Operator and Your Workspace

You can store intermediate results from calculations, as well as data and other things, by assigning them to an object with a name. The assignment operator looks like an arrow and is comprised of the less than sign followed by a dash, as in <-, or by a dash followed by a greater than sign, ->. Think about the assignment operator is as an arrow pointing to the object you want to assign something to, as in the following examples.

■ Example 1.4 — Illustrating how the assignment operator works.

```
> x <- 3                    # Assign x the value 3
> 27/3 -> y                 # Assign y the value 27/3 or 9
> z <- x + y                # Assign to z the sum of x + y
> vec1 <- c(1,2,3,4)        # Assign to vec1 a vector of integers
> vec2 <- c("A", "B", "C")  # Assign to vec2 a vector of characters
> x; y; z; vec1; vec2       # Print the variables to the screen
[1] 3
[1] 9
[1] 12
[1] 1 2 3 4
[1] "A" "B" "C"
```
■

In the example, the first and second lines show that assignments can be made in either direction (right to left or left to right). Note that the pound sign (#) denotes a comment and everything following it on that line is not processed by R. In the third line, we see that we can assign more than numbers; in this case we assign to z the sum of whatever is in x and y. In the fourth line we see that we can also assign a vector, in this case a vector consisting of the integers from 1 to 4, to an object called vec1. Here the vector is created using the concatenate function, c(). In the fifth line we see that we can also assign a character vector. In the sixth line of the example, note the use of the semi-colon as a line break. Here we could have typed x, y, z, vec1 and vec2 on separate lines, but the semi-colons allowed us to do it all on one line.

> **Exercise 1.3** Using R, reproduce the results of Example 1.4 on your computer. ∎

Now, we can also use the equals sign as an assignment operator. However, unlike with the "arrow" operators, it will only allow assignment from right to left, as the example below shows. In this text, we will not use the equals sign as an assignment operator.

■ **Example 1.5 — Using the equals sign as an assignment operator.**

```
> a = 7       # Assign a the value 7
> a           # Print the value of a to the screen
[1] 7
> 7 = b       # Attempting to assign to b the value 7 this way gives an error
Error in 7 = b :  invalid (do_set) left-hand side to assignment          ∎
```

Now, whenever you assign something like in the previous two examples, what you are actually doing is creating *objects* in your *workspace*. These objects can then be called by their names and used in all manner of ways that we will explore later. You can see what is in your workspace by using the list function, `ls()`. To simply list your workspace, the function does not require any arguments.

■ **Example 1.6 — Listing the workspace shows the objects we have created thus far.**

```
> ls()
[1] "a" "b" "vec1" "vec2" "x" "y" "z"                                      ∎
```

You can use the remove function, `rm(name)`, to delete the object "name" in your workspace. It works on single objects or a series of objects separated by commas.

■ **Example 1.7 — Removing an object from the workspace.**

```
> rm(a)
> ls()
[1] "b" "vec1" "vec2" "x" "y" "z"                                          ∎
```

Note that everything in your workspace only resides in active computer memory and not on your hard drive unless you expressly save it when you quit R. In other words, everything you do in an R session is temporary and will be lost when you exit R unless you tell R to save your work. See Section 1.3.5 for more discussion on this point.

"Reserved Words" and Special Characters

R has a number of what are called reserved words. The first is NaN, which stands for not a number. This usually arises when an improper mathematical operation is attempted, such as dividing by zero. R will then give NaN to indicate that it cannot give a numerical answer.

■ **Example 1.8 — An example of** NaN**.**

```
> 0/0
[1] NaN                                                                    ∎
```

Because these words are reserved you cannot use them as object names.

■ **Example 1.9 — An example of trying to use** NaN **as an object name.**

```
> NaN <- 3
Error in NaN <- 3 :  invalid (do_set) left-hand side to assignment        ∎
```

Also, infinity and minus infinity are denoted by the reserved words `Inf` and `-Inf` respectively.

■ Example 1.10 — Examples of `Inf` and `-Inf`.
```
> 1/0
[1] Inf
> -22/0
[1] -Inf                                                                           ■
```

As we will discuss in the next section, R is designed to work with vectors and, particularly with real-world data, a vector may sometimes be missing one or more components. A missing value is denoted by `NA`. As with `NaN`, `Inf`, and `-Inf`, you cannot use `NA` as an object name. Instead, you use it to tell R that a value is missing and, similarly, it will show up in R output to indicate when a value is missing.

■ Example 1.11 — Examples of `NA`.
```
> Y <- c(1,2,NA,4,5)
> Y
[1] 1 2 NA 4 5
> Z <- 5 + NA
> Z
[1] NA                                                                             ■
```

In the example, note how `NA` does not have quotes around it. Quotes, like in Example 1.4, indicate a character or character string. But as a reserved word, `NA` is not a character string. Also note how, as in the second part of the example, `NA` values are appropriately propagated through mathematical calculations. Here Z takes on the value `NA` since there is no way to know the value of `5 + NA`.

To test whether data are missing, use the is.na() function.

■ Example 1.12 — Testing for missing data.
```
> is.na(Y)
[1] FALSE FALSE TRUE FALSE FALSE
> is.na(Z)
[1] TRUE                                                                           ■
```

`NULL` is also a reserved word. It is often used to indicate that an object has no value assigned (which is not the same as a missing value denoted by `NA`).

Finally, R does not recognize smart quotes, which are the curvy quotes like this: " and ". These are often the default type of quotes used in many word processing programs. The alternative are plain quotes like this: " and ". If you use a word processor to save your R scripts, and it converts your quotes to smart quotes, you are likely to run into problems when you paste a command from the word processor into R.

■ Example 1.13 — What happens when smart quotes are used in an R command.
```
> X <- "a"
Error: unexpected input in "X <- ,"                                                ■
```

1.3.2 Some Important R Paradigms

The R design is based on a couple of paradigms that, once you know them, make it easier to understand how R works and thus it will be easier for you to work with R.

Objects

The first of these is that virtually everything in R is an *object*, of which there are various types. Each type of object has particular properties that control what an object can and cannot do, as well as how other objects interact with it.

Important types of objects in R include:

- List: an ordered collection of objects.
- Vector: a one-dimensional ordered collection consisting entirely of either numbers (integer, double, or complex), characters, or logicals.
- Array: a multi-dimensional vector consisting entirely of either numbers, characters, or logicals.
- Matrix: a two-dimensional array consisting entirely of either numbers, characters, or logicals.
- Data frame: a two-dimensional collection of data in which each of the columns can be of a different type (e.g., numeric, character, logical) and with rows of all the same length.
- Function: an R "program" that usually takes input as arguments and after running produces output.

You can learn about some of the properties of objects using the `typeof()` and `class()` functions.

■ **Example 1.14 — Learning a bit more about some previously created objects.**

```
> class(x)
[1] "numeric"
> typeof(x)
[1] "double"
> class(vec2)
[1] "character"
> typeof(vec2)
[1] "character"
```
                                                                                                                ■

The main point is that depending on the class and type of an object, R will or will not let you do certain operations. And note that when we previously created objects, we did not declare what the objects should be. R simply inferred what they should be from the input.

It is possible to coerce objects to be a different type and to test whether an object is of a particular type. For example, in Example 1.14 we learned that the object x is of type double. But we can change it to an integer using the `as.integer()` function and then test whether it is indeed an integer using the `is.integer()` function.

■ **Example 1.15 — Coercing an object's type.**

```
> typeof(x)
[1] "double"
> x <- as.integer(x)
> x
[1] 3
> typeof(x)
[1] "integer"
> is.integer(x)
[1] TRUE
```
                                                                                                                ■

Other functions for coercing include `as.numeric()`, `as.double()`, and `as.character()`. And, you can test objects with, for example, `is.numeric()`, `is.double()`, and `is.character()`, as well as `is.list()`, `is.vector()`, `is.matrix()`, `is.array()`, `is.data.frame()`, and `is.function()`.

Data frames

Data frames are the most frequently used R objects for storing and analyzing statistical data. Data frames have a specific structure: Columns are variables and rows are observations. In addition, a data frame must be rectangular, so all columns must be of equal length and all rows must be of equal length (though not necessarily of the same length as the columns). What is useful about data frames from an analysis point of view is that the columns can be of different types (for example, numeric, character, and logical).

The function data.frame() is used to create a data frame. The example below illustrates how to create a simple data frame.

■ **Example 1.16 — Creating a data frame.** First, we create some variables:

```
> First.name <- c("Joe", "Peggy", "Harry", "Joan")
> Last.name <- c("Sixpack", "Sue", "Henderson", "Jett")
> Age <- c(32, 27, 38, 35)
> Male <- c(TRUE, FALSE, TRUE, FALSE)
```

Note the use of the concatenate function, c(), which as we previously discussed is one way to create a vector of data. Next, assemble the variables into a data frame:

```
> practice.data <- data.frame(First.name, Last.name, Age, Male)
```

And, finally, print out the data frame to see what it looks like:

```
> practice.data
    First.name  Last.name  Age   Male
  1        Joe    Sixpack   32   TRUE
  2      Peggy        Sue   27  FALSE
  3      Harry  Henderson   38   TRUE
  4       Joan       Jett   35  FALSE
```

Note how the observations (rows) are numbered and the variables (columns) have the variable names at the top.   ■

Now that we have a data frame, we can extract individual variables using a $ notation. For example, if we wanted to use the Age variable from the practice.data data frame in a calculation, we would refer to it as practice.data$Age.

■ **Example 1.17 — Extracting variables from the practice.data data frame.**

```
> practice.data$Age
[1] 32 27 38 35
> practice.data$Male
[1] TRUE FALSE TRUE FALSE
```
                                                                                        ■

Exercise 1.4 Using R, reproduce the results of Examples 1.16 and 1.17 on your computer.   ■

Vector-based Calculations

R is very efficient at working with vectors and it is much less so when working with loops. If your previous programming experience is with a software language or languages in which, to iterate though a set of data, you had to use a loop, then programming in R will take some adjustment. In R, instead of writing code that operates on the rows of a data frame (i.e., the observations) you write code that operates on the columns (i.e., the variables).

■ **Example 1.18 — A simple vector calculation.** Let's start by creating a data frame:

```
> house.ID <- c("A", "B", "C", "D", "E")
> house.sq.ft <- c(1532, 2058, 1802, 1033, 1200)
> garage.sq.ft <- c(429, 836, 711, 0, 350)
> sale.price <- c(399900, 673000, 415000, 249900, 349900)
> house.data <- data.frame(house.ID, house.sq.ft, garage.sq.ft, sale.price)
```

Now to calculate the total square footage of the properties, it's as simple as:

```
> house.data$house.sq.ft + house.data$garage.sq.ft
[1] 1961 2894 2513 1033 1550
```

Notice that the vectors were added component-wise. This is also true when multiplying or dividing vectors (unless you use special operators to do matrix algebra).

Instead of outputting the result to the screen, you can also write it to the data frame:

```
> house.data$total.sq.ft <- house.data$house.sq.ft + house.data$garage.sq.ft
> house.data
    house.ID  house.sq.ft  garage.sq.ft  sale.price  total.sq.ft
1       A          1532         429         399900        1961
2       B          2058         836         673000        2894
3       C          1802         711         415000        2513
4       D          1033           0         249900        1033
5       E          1200         350         349900        1550
```

Here's the hard way to do the same calculation using a loop:

```
> house.data$total.sq.ft <- vector()
> for(i in 1:5){house.data$total.sq.ft <-
  house.data$house.sq.ft + house.data$garage.sq.ft}
> house.data$total.sq.ft
[1] 1961 2894 2513 1033 1550
```
■

Note the use of curly brackets in the second and third lines above. These are used to tell R that the input will take more than one line. While it will probably take some time to get used to thinking in terms of vectors rather than individual observations, once you do get used to it you'll find it's much easier to do analyses and to write code in R.

> **Exercise 1.5** Using the `house.data` data frame from Example 1.18, calculate the price per total square foot for each house.
>
> *Answer*:
> ```
> [1] 203.9266 232.5501 165.1413 241.9167 225.7419
> ```
> ■

1.3.3 Useful References and Getting Help

There are also a number online and installed references that you can access to learn more about R. A good, easily accessible (and free!) reference is *An Introduction to R* that was installed on your computer when you install R. You can access it via the pull-down menus:

• PC: Help > Manuals (in PDF) > *An Introduction to R*

- Mac: Help > R Help > Manuals > *An Introduction to R*

You can also get to the manuals online at `http://cran.r-project.org/manuals.html`. In addition, you might want to browse through the R Wiki at `http://rwiki.sciviews.org/doku.php` and perhaps the frequently asked questions (FAQs) at `http://cran.r-project.org/`.

There are many good reference books available, including:

- *The R Book* by Michael J. Crawley (published by Wiley).
- *R in a Nutshell: A Desktop Quick Reference* by Joseph Adler (published by O'Reilly).

Within R, if you know the name of the function you want to learn more about, you can look it up by using the `help()` function or by typing the name of function *without parentheses* preceded by a question mark. For example, if you wanted the help page for the quit function, on the command line you would either enter:

```
> help(q)
```

or

```
> ?q
```

This brings up the help page for the quit function. Try it and see what you get. When you do, note the standardized layout of the help pages where, at the very top left of the page the function name is displayed followed in the curly brackets by the package it comes from (we'll discuss packages shortly):

- **Description** tells you what the function does.
- **Usage** shows the syntax, including arguments and their default values.
- **Arguments** explains each of the function's arguments.
- **Details** provides additional description as necessary.
- **References** points the reader to additional documentation.
- **See Also** shows related functions. These are useful if the function you called up the help page for turns out to be not quite what you want.
- **Examples** at the bottom are little snippets of code you can copy and paste into R to see how the function works on a small example.

Of course, a difficulty with getting help this way is that you first need to know the name of the function. But quite often what you want to figure out is the name of the function itself. This is when the search engine is your friend! To search for something related to R, type the letter R (or R in square brackets, as in [R]) followed by what you want to learn about in the search engine box. For example, to learn about how to exit the R program, you might type [R] `how to quit` and, with just a bit of browsing through the search results, you will likely find what you are looking for.

Another useful trick for looking for help from within R is to use double question marks when you have an idea of what you are looking for but don't know the precise function name, as in `??topic.name`. This will return a list of all the help pages that contain "topic.name" and from which you can then try to narrow down your search. Equivalently, you can also type `help.search("topic.name")`; note the quotes around the search term which are required.

> **Exercise 1.6** What do you get when you type:
> ```
> > ??quit
> ```
> ∎

1.3.4 Extending R: Installing Packages

When you installed R on your computer, you installed the "base package." One of the strengths of R is that its capabilities can be expanded by installing additional packages (new ones of which are being written and expanded on all the time). So, let's learn how to download and install these additional packages.

A package we are going to use later in the text is called "bootstrap." The PC and Mac R console interfaces are slightly different in terms of how to download and install R packages, so follow the appropriate directions below.

PCs using the Console:
- To download it, first make sure your computer is connected to the internet and then click on the "Packages" pull-down menu at the top of the R Console window.
- Scroll down to and click on "Install package(s)..."
- When the "CRAN mirror" window pops up, pick a site geographically near your location and click OK.
- When the "Packages" window opens, scroll down to "bootstrap" (on the way, look at all the other packages!), highlight it, and click OK.
- You should get a progress bar and some output in the R Console window that, if all went well, somewhere in the middle says "package 'bootstrap' successfully unpacked and MD5 sums checked."
- If you get an error and are running Windows Vista, you may first have to start R with administrator privileges.

Macs using the Console:
- To download it, first make sure your computer is connected to the internet and then click on the "Packages & Data" pull-down menu at the top of the screen.
- Scroll down to and click on "R Package Installer."
- Click on "Get List" and pick a site geographically near your location and click OK. (Say yes to keeping the site as the default if asked.)
  - Alternatively, you can type "bootstrap" in the Package Search box. This will bring up all packages with the word bootstrap in their titles.
- When the "Package" screen fills up, scroll down to "bootstrap" (on the way, look at all the other packages!), highlight it, be sure the "Install Dependencies" box is checked, and then click "Install Selected."
- You should get a progress bar and some output in the R Console window that, if all went well, somewhere says "The downloaded packages are in..."

Exercise 1.7 Install the `bootstrap` package onto your computer. ∎

To see what packages that are currently available on your system type:
```
> library()
```

To load a package, say bootstrap, so that you can access the data or functions, use the `library()` function, as in
```
> library(bootstrap)
```

Alternative for Macs: Packages & Data > Package Manager. Then check the box next to bootstrap (or whatever package you want activated).

To detach a package when you are done, again say bootstrap, type

```
> detach("package:bootstrap")
```

where you must include the quotes in the expression.

1.3.5 A Few Final Notes

Before we wrap up this brief introduction to R, just a couple of more notes. First, Appendix A provides additional information about R, including how to write scripts and functions in R. While these topics are not necessary for the R examples and exercises throughout the book, they come in very handy for doing actual statistical analyses with R.

Also, statistical analyses require data. In the next section we will briefly discuss how to read data into R – enough to get you started. For those who want more details, see Appendix A and the references in Section 1.3.3. And, last but certainly not least, we will finish by learning how to quit R.

Reading Data into R

In previous examples in this chapter, for example Examples 1.16 and 1.18, we've seen that we can use the assignment operator to input data into R. The scan() function is also useful for inputting small sets of data directly.

■ Example 1.19 — Using the scan() function to input data.

```
> some.data <- scan()
1:  43
2:  27
3:  8
4:  95
5:
Read 4 items
> some.data
[1] 43 27 8 95
```

Note how in the first line assigning the scan() function to an object puts R into a data entry mode (denoted by the change in command prompt in the subsequent lines). Then all you do is type each data item followed by the ENTER or RETURN key. Continue doing this until you are out of data at which time hit the ENTER or RETURN key on an empty line. This tells R you're done.   ■

Of course, these methods are only practical for very small data sets. If the data you need to enter already exists in electronic form, the easy way to enter it into R is as a comma-separated values (csv) file. Whatever software your data resides in can usually be saved in csv format.

Given your data are in csv format, then it can be read using the read.csv() function, where the argument is the file name (including the appropriate path). For example, if a csv data file called data.csv is located on a Mac computer with path /Users/rdfricker/Desktop, it could be read in to R and to a data frame object called my.data by typing:

```
> my.data <- read.csv("/Users/rdfricker/Desktop/data.csv")
```

If you are a PC user, note the use of backslashes rather than forward slashes. Also note that the default option for the `read.csv()` function is that the first line of the file contains the variable names and then each subsequent line contains the data, one line for each observation (row) in the data, where data item is separated from the next by a comma.

Now, explicitly specifying the path to the file in the `read.csv()` function is tedious and prone to error (and can be particularly confusing if you are working on a PC where the paths are specified using forward slashes). However, unless you are writing scripts (see Appendix A) or otherwise need to automate reading in data, there is an easier way using the `file.choose()` function. Specifically, if you type the following command

```
> my.data <- read.csv(file.choose())
```

the `file.choose()` function will first open up a dialog box on your computer from which you can simply find and open the desired file in the usual way and, once you select it, the `read.csv()` function will then read it in. Give it a try. This is the method we will use throughout the text for reading in data.

Functions

One of the strengths of R is the ability to use and write functions. Functions are basically mini-programs, where you can create a function that even calls other functions. Just about everything you do in R involves applying a function, usually to data. And, as we'll discuss shortly, R users can write their own functions that can be published to the wider R community via packages that everyone can then download and use.

R has thousands of functions. We'll get to many of them as we go through the text. For now, below is a list of those you need to know to get started.

- `c()` is the concatenate function, which is often used to create a vector, as in `vector1 <- c(1,2,3)` or to join two or more vectors, as in `new.vector <- c(vector1,vector1)`.
- `detach("package:`*name*`")` detaches the package called "name."
- `dim(`*data*`)` returns the number of rows and columns respectively in "data," which can be either a data frame object or matrix object.
- `help(`*function.name*`)`, `?`*function.name*, and `??`*text* are useful for getting help.
- `is.na()` is a function that returns a logical object indicating whether data are missing (i.e., NA) or not.
- `length(`*vector*`)` returns the number of elements in "vector."
- `library(`*name*`)` loads the package called "name" so you can use/access its contents.
- `ls()` lists the object in the workspace. Note that no arguments are required to run the function, but you do have to type the parenthesis because it is a function.
- `read.csv()` and `read.table()` are useful for reading data into R. For reading data in in specialized formats, the `foreign` package is very useful.
- `rm(`*name*`)` deletes the object "name" from the workspace. It works on single objects or a series of objects separated by commas.
- `rm(list = ls(all = TRUE))` deletes all the files in your workspace. Be very cautious when using this – there is no undo option!
- `q()` quits the current R session. Like `ls()`, there are no arguments to the function, but you have to type the parenthesis.

For most of the text, we will focus on the application of existing R functions for statistical data analysis. However, if you would like to learn more about writing functions, see Appendix A and the references listed in the next section.

Quitting R

Finally, besides knowing how to start R, you also need to know how to quit R. On most GUIs, like the Mac version shown in Figure 1.3, there is an icon as well as a menu item for quitting R. But you can also quit R by typing q(), another function, at the command prompt, where you must type the parentheses after the letter q.[3]

Now, when you tell R to quit, it will ask you whether you want to save the session (or, equivalently, whether you want to save the workspace image). If you choose "No," then all the work you have done in that session will be deleted. Be careful when exiting because once you say "No" to saving your session all the work from that session is lost.

Of course, you can say "Yes" to saving the session and that does two things. First, it will save your workspace so that, when you return to R, all the objects will be available. Second, it will also save your work history (i.e., all the commands you ran) so that they are also available (via the up and down arrow keys) in the future session.

Now, it might seem like the best option is just to save every session so nothing is lost. The problem with that is lots of unwanted stuff will build up in your workspace making it hard to find the objects you want. A good strategy is to first remove any unwanted files (using the rm() function) then quit R where, when R asks "Save workspace image?"[4], say yes.

---

[3] If you type the name of a function into R without parenthesis, R will print the function's code to the screen. Once you become adept at R syntax, this can be useful for understanding precisely how a function works. For example, if you type q without parentheses you will see the code for the quit function rather than quitting R Try it.

[4] See Appendix A for additional workspace management tips.

1.4 Problems

Problem 1.1 In your own words, explain why statistics is important to the practice of data science.

Problem 1.2 Find an example, either in the popular media or on the internet, of statistics being used in a data science application or problem.

Problem 1.3 In your own words, describe how statistics and data science are similar and how they differ.

Problem 1.4 Discuss the four ways that statistics can be used to address data science questions and give an example of each.

Problem 1.5 In your own words, explain the difference between descriptive and inferential statistics.

Problem 1.6 Either in the popular media or on the internet, find an example of a data science application that is using statistics to solve or help solve the problem and then find another data science application that *should be* using statistics to solve or help solve the problem but is not.

Problem 1.7 Using R, calculate the following quantities:

a. $(37.923 - 14.720) * 2.5$
b. $\left(\sqrt{833} + 47.5\right) / 275.8^2$
c. $\log_{10}(50)$
d. $\ln(50)$

Turn in both your R code and the solutions. Note that the log() function in R calculates the natural log and the log10() function in R calculates log base 10.

Problem 1.8 Using R, calculate the following quantities:

a. $(82.71 + 2.73) / (7.32 \times 9.58)$
b. $\frac{72}{195} \times \frac{33}{94.7}$
c. $\cos(20)$
d. e^2

Turn in both your R code and the solutions. Note that the cos() function in R calculates the cosine and the exp(x) function in R calculates e^x.

Problem 1.9 Using the practice.data data frame from Example 1.16, in R, calculate the age of the individuals in months. Turn in both your R code and the solutions.

Problem 1.10 Using the house.data data frame from Example 1.18, in R, calculate the price per *house* square foot of each home. Do the calculation using the appropriate vectors from the data frame. Turn in both your R code and the solutions.

Problem 1.11 Look up the R help page for the read.table() function. What is the difference between the read.csv() and read.table() functions?

Problem 1.12 If you didn't complete Exercise 1.7, install the bootstrap package. Also install the shiny, lattice, and wordcloud packages on your computer.

2 — Descriptive Statistics

2.1 Introduction

Descriptive statistics is all about summarizing data. In this chapter we will learn how to describe data numerically using statistics and in Chapter 3 we will learn how to graphically describe data. In fact, a *statistic* is simply a number calculated from data that summarizes something about the data. For example, the average is a statistic, and there are many other types that we will learn about in this chapter. So, a statistic is just a number. That number may be more or less informative about the whole data set. In comparison, a *descriptive statistic* should be just that: It should usefully describe a set of data.

Descriptive statistics are becoming increasingly important as data scientists deal with ever larger data sets and as data collection accelerates in our ever more computerized and interconnected world. They are important because the human mind is as limited as it ever was for assimilating individual facts – we simply aren't good at being able to synthesize lots of numbers. Indeed, human short-term memory capacity is only about seven digits – the length of U.S. telephone number *not* counting the area code. So, in our increasingly data-rich world, knowing how to appropriately summarize data is a critical skill.

For example, returning to our discussion of the U.S. Census in Chapter 1, the only way to get some understanding of the United States population using the Census is to use descriptive statistics (as well as other methods we will talk about in later chapters) to summarize the data. Looking at individual Census records would not tell us much about the entire U.S. population. And, these days, the 300 million plus Census records is not a big data set. There are more tweets sent *per day* than there are Census records. The only way to gain insight into these massive data sets is to appropriately summarize them so that the human mind can assimilate and interpret the information contained in the data.

Before we proceed further, let's formalize what we've just discussed with some definitions.

Definition 2.1.1 — Data. Information, often numerical but not necessarily so, collected from an experiment, a survey, administrative records, the Internet, etc. The word "data" is plural. One piece of information is a "datum."

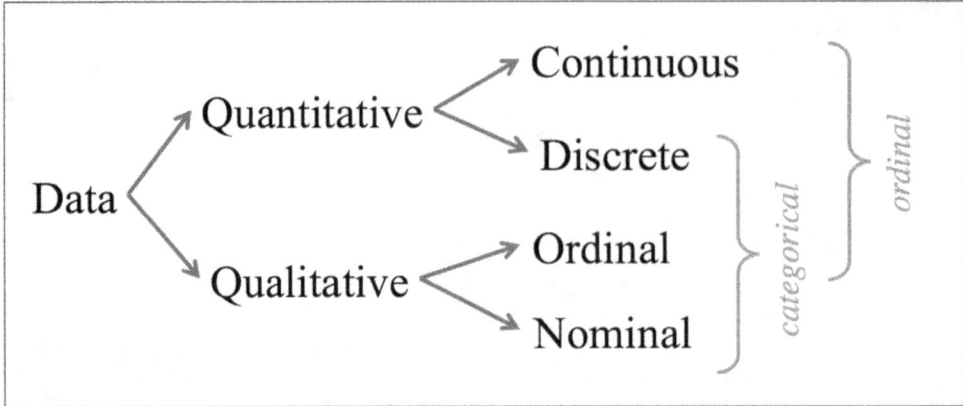

Figure 2.1: A taxonomy of types of data.

Definition 2.1.2 — Statistic. A numerical fact, usually computed from a data set. Statistics can also be computed from subsets of the data and can even be just a single datum.

Definition 2.1.3 — Descriptive Statistic. A statistic that usefully summarizes a data set, where the data can be either for an entire population or for a subset of the population.

Good descriptive statistics can help data scientists understand what the data are trying to say. They can highlight and bring out the underlying information in a data set, which might not (and probably will not) be evident by just inspecting the individual data elements. .

2.1.1 Types of Data

Before proceeding further, it is important to understand that not all data are alike. Most fundamentally we can divide data into two basic types, quantitative and qualitative. *Quantitative* data are data that can be measured or characterized with a numerical value while *qualitative* data cannot be so measured. For example, if we think about demographic data, height, weight, and age are all quantitative, while gender and eye color are qualitative.

We can also divide data into *cross-sectional* data and *longitudinal* (also known as *time series*) data. Cross-sectional data are data that occur either in one time period or are constant over time, while longitudinal data occurs over multiple time periods or varies over time. For example, referring back to the previous demographic data, gender and eye color are cross-sectional in the sense that they are unlikely to change over time. On the other hand, height, weight, and age data could be cross-sectional if they are recorded for one period of time and longitudinal if they are repeatedly measured over multiple time periods.

As shown in Figure 2.1, we can further describe quantitative data as either *continuous* or *discrete*. Data are discrete if there are gaps between the values the data can assume. For example, the number of people in a family can be 1, 2, 3, ... It cannot be 2.7. If there are no gaps between possible data values, then we say the data are continuous. Another way to think about continuous data is that if we had an infinitely accurate measuring device we could express the data to any number of decimal places and it would always make sense. So, for example, height is continuous: we can talk about someone being 6 feet tall, or 5.97 feet tall, or 5.9722683 feet tall.

Unlike quantitative data, qualitative data cannot be measured or described numerically. As shown in Figure 2.1, qualitative data can be either *nominal* or *ordinal*. Ordinal qualitative data are data for which there is a natural ordering, but the data cannot be expressed numerically. For example, shirt size is ordinal: "large" shirts are bigger than "medium" shirts which are bigger than "small" shirts. In contrast, with nominal data there is no natural ordering to the data. For example, gender is a nominal type of data: each person can be classified as "male" or "female," but it does not make any sense to say that "male" is greater than "female."

Note that nominal data can be represented numerically for purposes of analysis, but care must be taken not to over interpret the numerical labels. For example, we will sometimes use an *indicator variable* to do analyses, say setting a variable called gender equal to 1 for men and 0 for women. But just because the numbers 0 and 1 are ordinal, this property does not carry over to the original qualitative variable, and so care must be taken not to over interpret or misuse the indicator variable.

We use the term *categorical* to refer to data that is either discrete or qualitative because we can naturally categorize these types of data into groups. And note that continuous data are trivially ordinal since numeric data has an obvious ordering. Finally, note that we can turn continuous data into categorical data by defining ranges of values for each category. For example, for height, people may be categorized as "short" if they are less than five feet tall; "average" if they are between five and six feet; and "tall" if they are six feet or greater in height.

The reason these distinctions are important is that the appropriate statistical analyses, and even the proper way to display data, will depend on the type of data.

2.1.2 Example Data: U.S. Domestic Flights from 1987 to 2008

To help make the ideas and methods of this chapter concrete, we will illustrate them using a data set consisting of United States domestic airline flight arrivals and departures and associated details for all commercial flights from October 1987 to April 2008. This is a large dataset: there are nearly 120 million records for 3,376 airports. In this chapter, we'll mainly focus on the data from years 1988, 1997, and 2007, where there are 5,202,096, 5,411,843, and 7,453,215 observations (i.e., flights), respectively, for those years.

The airline data are available at `http://stat-computing.org/dataexpo/2009/`. As shown in Table 2.1, the data set contains information on the date, day, and time of each flight, the airline and particular airplane, flight origin and destination, and lots of information about the length of each flight and the types of delays, if any, experienced. Table 2.2 shows five randomly-selected observations from the 2007 data.

Looking through Table 2.2 we see, for example, that the time an airplane is in the air (see the *AirTime* variable) and the distance flown (see the *Distance* variable) are continuous data while the day of the month (see the *DayofMonth* variable) is an example of discrete data. On the other hand, the reason for flight cancellation (see the *CancellationCode* variable) and airport of origination (see the *Origin* variable) are examples of nominal data, while the flight month (see the *Month* variable) is an example of ordinal data.

It is worth noting that some of the data that are numerically coded are actually qualitative. For example, as we see in Table 2.2, day of the week (see the *DayOfWeek* variable) contains the integers from 1 to 7, where the number 1 corresponds to Monday, 2 to Tuesday, etc. Though these data are numbers, they are not quantitative data *per se*. Instead, the numbers are only codes that represent ordinal qualitative data. Similarly, note how the flight numbers (*FlightNum* variable) are also coded numerically, but they are really nominal data since the order of the numbers does not have any meaning.

Table 2.1: Aircraft data set variables and brief descriptions. Additional information on the variables is available at `http://www.transtats.bts.gov/Fields.asp?Table_ID=236`.

	Variable Name	**Description**
1	Year	1987-2008
2	Month	1-12
3	DayofMonth	1-31
4	DayOfWeek	1 (Monday) - 7 (Sunday)
5	DepTime	actual departure time (local, hhmm)
6	CRSDepTime	scheduled departure time (local, hhmm)
7	ArrTime	actual arrival time (local, hhmm)
8	CRSArrTime	scheduled arrival time (local, hhmm)
9	UniqueCarrier	unique carrier code
10	FlightNum	flight number
11	TailNum	plane tail number
12	ActualElapsedTime	in minutes
13	CRSElapsedTime	in minutes
14	AirTime	in minutes
15	ArrDelay	arrival delay, in minutes
16	DepDelay	departure delay, in minutes
17	Origin	origin IATA airport code
18	Dest	destination IATA airport code
19	Distance	in miles
20	TaxiIn	taxi in time, in minutes
21	TaxiOut	taxi out time in minutes
22	Cancelled	1 = yes, 0 = no
23	CancellationCode	reason for cancellation (A = carrier, B = weather, C = NAS, D = security)
24	Diverted	1 = yes, 0 = no
25	CarrierDelay	in minutes
26	WeatherDelay	in minutes
27	NASDelay	in minutes
28	SecurityDelay	in minutes
29	LateAircraftDelay	in minutes

Table 2.2: Five randomly-selected observations from the 2007 aircraft data set.

	Year	Month	DayofMonth	DayOfWeek	DepTime
1	2007	3	30	5	1339
2	2007	4	2	1	NA
3	2007	4	30	1	1353
4	2007	12	27	4	2235
5	2007	1	20	6	805

	CRSDepTime	ArrTime	CRSArrTime	UniqueCarrier	FlightNum
1	1340	1548	1588	YV	7352
2	824	NA	1005	OO	6625
3	1348	1420	1410	AS	381
4	2135	143	33	B6	861
5	805	1045	1033	YV	2681

	TailNum	ActualElapsedTime	CRSElapsedTime	AirTime	ArrDelay
1	N516LR	69	78	49	-10
2	0	NA	101	NA	NA
3	N788AS	87	82	63	10
4	N588JB	188	178	146	70
5	N911FJ	160	148	139	12

	DepDelay	Origin	Dest	Distance	TaxiIn
1	-1	ORD	CAK	344	3
2	NA	DEN	BZN	525	0
3	5	BOI	SEA	399	8
4	60	JFK	TPA	1005	5
5	0	PWM	CLT	812	8

	TaxiOut	Cancelled	CancellationCode	Diverted	CarrierDelay
1	17	0		0	0
2	0	1	A	0	0
3	16	0		0	0
4	37	0		0	0
5	13	0		0	0

	WeatherDelay	NASDelay	SecurityDelay	LateAircraftDelay
1	0	0	0	0
2	0	0	0	0
3	0	0	0	0
4	0	10	0	60
5	0	0	0	0

2.2 Cross-sectional Data

As we just discussed, cross-sectional data are collected during the same period of time. Statistics may then then be used to summarize this data.

2.2.1 Measures of Location

Measures of location, sometimes also referred to as *measures of central tendency*, are typically used to quantify where the "center" or mass of the data are located. The word center is in quotes as there are a number of common measures of central tendency, each of which quantifies the "center" in a different way. The most common measure is the *mean* which is the average of a set of observations in either a sample or a population.

The average is something most people are familiar with. It is simply the sum of all the data ("observations") in either the sample or the population divided by the total number of observations in the sample or population. For example, if we have data on an entire population of size N, then the population mean is calculated as shown in Definition 2.2.1.

> **Definition 2.2.1 — Population mean.** For data from a population, denoted $x_1, \ldots, x_N$, the population mean is calculated as
>
> $$\mu = \frac{1}{N} \sum_{i=1}^{N} x_i = \frac{x_1 + x_2 + \cdots + x_n}{N}.$$

The population mean is denoted by the Greek letter μ, which is pronounced "mu," and the individual observations (i.e., the data) are denoted by x_i for $i = 1, 2, 3 \ldots, N$.

If we only have data on a sample of the population, where we denote the sample size by n, $n < N$, then the *sample average* or *sample mean*, is the sum of the sample data divided by n.

> **Definition 2.2.2 — Sample mean.** For a sample of data, $x_1, \ldots, x_n$, the sample mean $\bar{x}$ is calculated as
>
> $$\bar{x} = \frac{1}{n} \sum_{i=1}^{n} x_i.$$

As shown in the definition, the sample mean is denoted by $\bar{x}$. This is spoken as "x-bar," which obviously comes from the notation: an x with a bar over it.

■ **Example 2.1 — Calculating the sample mean.** Calculate the mean for the following sample of data: $\{2.3, 8.1, 5.5, 9.0, 7.8\}$.

Solution: Here we have $n = 5$, so

$$\bar{x} = \frac{1}{5} \sum_{i=1}^{5} x_i = \frac{2.3 + 8.1 + 5.5 + 9.0 + 7.8}{5} = \frac{32.7}{5} = 6.54.$$

                                                                           ■

A couple of additional points on notation:

- Using lower case Roman letters for data (i.e., x) indicates that this is a calculation for a particular sample of numbers. That is, the letter x represents a particular number. (Beginning in Chapter 4, we will talk about random variables which will be denoted with capital Roman letters.)

- Similarly, it is convention to use the letter n to indicate the number of items in a sample and for N to denote the population size. Assuming the sample is drawn from the population *without replacement*, meaning each observation in the population can only be drawn and put into the sample once, then for any given data set $n \leq N$.
- In later chapters we will use the sample mean to estimate the population mean (for those times when it is not possible to observe the whole population). In such cases it is common to use the notation $\hat{\mu}$ (spoken "mu hat") instead of $\bar{x}$ to indicate that the sample mean is an estimate of the population mean μ. (This notation is used more generally, putting a "hat" over the symbol for a population quantity to indicate an estimate of that population quantity.)

In R, the `mean()` function can be used to calculate either a population or sample mean depending simply on whether the data set is for a population or a sample. For example,

```
> mean(c(2.3, 8.1, 5.5, 9.0, 7.8))
[1] 6.54
```

Of course, the real power of R comes not from using it to calculate the mean of 5 numbers – we can do that easily on a calculator or in Excel – but in using it to do calculations for larger data sets such as the airline data. For example, let's look at airline departure delays (contained in the *DepDelay* variable), which is defined as the difference in minutes between the scheduled and actual departure times. Example 2.2 shows that the average departure delay for all flights, and it shows that the average departure delay has been increasing over the past 20 years.

■ **Example 2.2 — Using R to calculating the mean.**
```
> data88 <- read.csv(file.choose()) # read in 1988.csv
> data97 <- read.csv(file.choose()) # read in 1997.csv
> data07 <- read.csv(file.choose()) # read in 2007.csv
> mean(data88$DepDelay, na.rm=TRUE)
[1] 6.706768
> mean(data97$DepDelay, na.rm=TRUE)
[1] 8.235566
> mean(data07$DepDelay, na.rm=TRUE)
[1] 11.39914
```
■

In Example 2.2, setting the `na.rm` option to TRUE tells the mean function to ignore all the missing observations. Thus, the example shows that the mean departure delay *for those flights that had data* was 6.7 minutes in 1988, 8.2 minutes in 1997, and 11.4 minutes in 2007. This raises the question of the fraction of flights for which we're missing data in each year. Example 2.3 calculates the fraction of flights for which data are missing (i.e., *DepDelay* is NA) by calculating the the mean of a missing data indicator variable.

■ **Example 2.3 — Missing data fraction calculated via the mean of an indicator variable.**
```
> miss_ind_88 <- as.numeric(is.na(data88$DepDelay))
> mean(miss_ind_88)
[1] 0.009642844
> mean(as.numeric(is.na(data97$DepDelay)))
[1] 0.01806464
> mean(as.numeric(is.na(data07$DepDelay)))
```

```
[1] 0.02156761
```
∎

Example 2.3 shows that in 1987 just under one percent of the flights did not have departure delay data. This increased to just under two percent in 1997 and just over two percent in 2007. Thus, overall, the amount of missing data are small.

To understand the R code in Example 2.3, first note that the `is.na()` function generates a vector of true and false values, where the value is true whenever *DepDelay* is NA. Then the `as.numeric()` function turns the true values into ones and the false values into zeros so that *miss_ind_88* is an indicator variable that takes on the value 1 if *DepDelay* is NA and 0 otherwise. Then calculating the mean of this indicator variable gives the fraction of flights that are missing flight delay data.

Let's confirm that this is correct by using the `table()` function to actually count the number of true and false values and, equivalently, the number of one and zero values in the indicator variable for the 1988 data:

∎ **Example 2.4 — Checking the calculations for 1988 in Example 2.3.**

```
> table(is.na(data88$DepDelay)) # count the number of NAs
  FALSE   TRUE
5151933 50163
> table(miss_ind_88) # count the number of 1s and 0s in the indicator variable
      0     1
5151933 50163
> 50163/(5151933+50163) # now calculate the fraction of ones - it matches
[1] 0.009642844
```
∎

Another measure of central tendency is the *median*. The median is simply the middle value of the <u>ordered</u> data. If there are an odd number of observations, it is literally the value of the middle observation of the ordered data. If there is an even number of observations, then it is the average of the middle two points of the ordered data.

To express the median in a mathematical formula, we will use the notation $x_{(i)}$ which denotes the ith ordered observation. For example, returning to the data from Example 2.1, $x_{(1)} = 2.3$, $x_{(2)} = 5.5$, $x_{(3)} = 7.8$, $x_{(4)} = 8.1$, and $x_{(5)} = 9.0$. Then, using this notation, the median is defined below.

Definition 2.2.3 — Median. For a set of data, denoted $x_1, \ldots, x_n$, the median $\tilde{x}$ is calculated as

$$\tilde{x} = \begin{cases} x_{\left(\frac{n+1}{2}\right)}, & \text{if } n \text{ is odd} \\ \left[x_{\left(\frac{n}{2}\right)} + x_{\left(\frac{n}{2}+1\right)} \right] \big/ 2, & \text{if } n \text{ is even.} \end{cases}$$

For the median, $\tilde{x}$ is spoken as "*x*-tilde."

∎ **Example 2.5 — Calculating the sample median.** Calculate the median for the data from Example 2.1: {2.3, 8.1, 5.5, 9.0, 7.8}.

Solution: Because $n = 5$ is odd, $\tilde{x} = x_{\left(\frac{n+1}{2}\right)} = x_{\left(\frac{5+1}{2}\right)} = x_{(3)} = 7.8$. ∎

In R, the `median()` function can be used to calculate the median. For example,

```
> median(c(2.3, 8.1, 5.5, 9.0, 7.8))
[1] 7.8
```

While both the median and the mean are measures of central tendency, they are distinctly different. If, when we use the term central tendency, we mean a typical value in the middle of the data, then it often makes sense to use the median. If, on the other hand, we really want to know the average, then the mean is the appropriate measure. When the data contain one or more *outliers*, meaning data points that are unusually large or small when compared to the rest of the data, the median may be preferred to the mean. This is because, as the next example shows, the median is less affected by outliers.

■ **Example 2.6** Consider a sample of seven observations: $\{0,1,2,2,2,3,4\}$. For this data, the mean equals the median:

$$\bar{x} = \frac{1}{7}\sum_{1}^{7} x_i = \frac{0+1+2+2+2+3+4}{7} = \frac{14}{7} = 2,$$

and

$$\tilde{x} = x_{\left(\frac{7+1}{2}\right)} = x_{(4)} = 2.$$

Yet, what if a mistake was made entering the data so that the first observation was accidentally recorded as 70 instead of 0? In that case, the median would be unchanged at 2, but the mean would be 12. This is clearly a big difference, and it's only because of a typographical error for one datum. ■

> **Exercise 2.1 — Calculating means and medians.** Confirm that for the data with the typo, $\{70,1,2,2,2,3,4\}$, $\bar{x} = 12$ and $\tilde{x} = 2$. ■

If the data have one or more large outliers, then the use of the mean as a measure of central tendency could be questionable. For example, with the mistaken data set from Exercise 2.1, the resulting sample average is larger than all but one of the observations. However, the median is the same whether the outlier is included in the data or not. One way to describe this property is to say that the median is *robust* to outliers while the mean is sensitive to them. Whenever data are very *skewed*, meaning some observations are significantly larger or smaller than most of data, then the median is likely to be a better measure of central tendency since it better describes a "typical" value in the data.

To bring this idea home, imagine we were trying to describe the typical net worth of a new graduate with a master's degree in data science. Since starting salaries tend to be similar in magnitude, using the mean or the median is likely to give about the same picture. However, if Bill Gates had become bored with running his foundation and decided to earn a data science degree, using the average to describe a typical new master's net worth might be just a bit misleading!

Example 2.7 calculates the median flight departure delay, where here we see that for all three years it was 0. Thus, in spite of the fact that the average delay is increasing, for each year at least 50 percent of the flights had no delay.

■ **Example 2.7 — Using R to calculating the median.**
```
> median(data88$DepDelay, na.rm=TRUE)
[1] 0
> median(data97$DepDelay, na.rm=TRUE)
[1] 0
> median(data07$DepDelay, na.rm=TRUE)
[1] 0
```
■

The *trimmed mean* is another measure of central tendency that can be a useful compromise between the mean and the median. For $0 \leq p \leq 0.5$, the $100p$ percent trimmed mean is calculated by first discarding the $100p$ percent largest and smallest observations in the data and then averaging the remaining observations.

Mathematically, let $x_{(1)} \leq x_{(2)} \leq \ldots \leq x_{(n)}$ be the order statistics of n observations. Then the trimmed mean $\bar{x}_{tr(p)}$ is defined below.

> **Definition 2.2.4 — Trimmed mean.** For a sample of data, $x_1, \ldots, x_n$, and some specified fraction $0 \leq p \leq 0.5$, the $100p$ percent trimmed mean is calculated as
>
> $$\bar{x}_{tr(p)} = \frac{1}{n-2m} \sum_{i=m+1}^{n-m} x_{(i)},$$
>
> where $m = p \times n$. If m is not an integer, simply round it down to the nearest integer.

Note that the $100p$ percent trimmed mean actually deletes $2 \times 100p$ percent of the data. Also note that $\bar{x}_{tr(0)} \equiv \bar{x}$ and $\bar{x}_{tr(0.5)} \equiv \tilde{x}$, though for any other value of $0 < p < 0.5$, there is no guarantee that the trimmed mean will be between the mean and median.

■ **Example 2.8 — Calculating and comparing trimmed means.** Calculate the 20 percent trimmed mean for the data from Example 2.6, both with and without the typo.

Solution: First, because $p \times n = 0.2 \times 7 = 1.4$, $m = 1$. Then the original set of ordered data $\{0, 1, 2, 2, 2, 3, 4\}$ is trimmed to $\{1, 2, 2, 2, 3\}$. That is,

$$\bar{x}_{tr(0.2)} = \frac{1}{n-2m} \sum_{i=m+1}^{n-m} x_{(i)} = \frac{1}{5} \sum_{i=2}^{6} x_{(i)} = \frac{1+2+2+2+3}{5} = \frac{10}{5} = 2.0.$$

In a similar way, the data with the typo, $\{70, 1, 2, 2, 2, 3, 4\}$, is trimmed to $\{2, 2, 2, 3, 4\}$ so that

$$\bar{x}_{tr(0.2)} = \frac{2+2+2+3+4}{5} = \frac{13}{5} = 2.6,$$

which, while somewhat larger than $\tilde{x} = 2$, is not nearly as large as $\bar{x} = 12$.   ■

In R, the trimmed mean is calculated using the `mean()` function with the `trim` option. Example 2.9 shows that R gives the same answer as the manual calculation in Example 2.8.

■ **Example 2.9 — Using R to calculate the trimmed mean.**

```
> mean(c(0,1,2,2,2,3,4),trim=0.2)
[1] 2.0
> mean(c(70,1,2,2,2,3,4),trim=0.2)
[1] 2.6
```
                                                                                                      ■

Note that the mean, median, and trimmed mean are all appropriate to use with either continuous or discrete data. Returning to the previous example of children in a family, while it does not make sense to say that any particular family has 2-1/2 children, because of the discrete nature of the data, it is perfectly fine to say that the average number of children per family is 2.5. However, when data are nominal, none of these measures work. For nominal data, the *mode*, which is the category with the largest number of observations, is the appropriate measure.

For example, in the airline delay data, we previously noted that in spite of it being coded numerically, the *DayOfWeek* variable contains qualitative (ordinal) data. Because it's qualitative, it would not make sense to calculate the mean or median for *DayOfWeek*. But, as shown in Example 2.10, the mode tells us which day of the week has the most flights: for 1988, it's Friday (*DayOfWeek* = 5); for 1997, it's Wednesday (*DayOfWeek* = 3), and for 2007 it's Monday (*DayOfWeek* = 1).

■ **Example 2.10 — Using R to calculate the mode.**

```
> table(data88$DayOfWeek)
     1      2      3      4      5      6      7
755898 757140 757963 753415 766364 697795 713521
> table(data97$DayOfWeek)
     1      2      3      4      5      6      7
790298 791617 802130 785731 786342 706198 749527
> table(data07$DayOfWeek)
      1       2       3       4       5      6       7
1112474 1078562 1088858 1097738 1101689 933338 1040556
```
■

Note that the R mode() function displays the storage mode of an R object, it does not calculate the mode of a set of data. Also, the term mode has a different connotation for numeric data. We'll get to that in later chapters when we talk about distributions.

> **Exercise 2.2** Both manually and in R, calculate the mean, median, and 10 percent trimmed mean of the following sample of data: {98.6, 92.4, 97.4, 118.6, 84.1, 111.1, 84.1, 103.8, 90.5, 102.4, 101.0, 100.6, 109.6, 102.5, 106.0, 98.7, 79.5, 85.0, 114.1, 89.3}.
>
> *Answer*: $\bar{x} = 98.465$, $\tilde{x} = 99.65$, and $\bar{x}_{tr(0.1)} = 98.3125$. ■

2.2.2 Measures of Variation

When describing or summarizing a set of data, providing measures of both location and variation are important. That is, while the mean, trimmed mean, and median provide information about the average or typical observation in the data, they do not give any information about the variability of the data. But knowing whether the data tend to be close together or spread far apart can be useful and informative. This section presents useful numerical summary *measures of variation* and Section 3.2 of the next chapter presents graphical plots that allow one to visually assess data variability.

The most common measure is the standard deviation. It is based on the *sample variance*, s^2, which is defined below.

> **Definition 2.2.5 — Sample variance.** For a sample of data, $x_1, \ldots, x_n$, the sample variance s^2 is calculated as
>
> $$s^2 = \frac{1}{n-1} \sum_{i=1}^{n} (x_i - \bar{x})^2.$$

The formula for the sample variance in Definition 2.2.5 is a bit more complicated than the mean Definitions 2.2.1 or 2.2.2. Starting within the parenthesis, the formula says to take the difference between each observation in the sample and the sample mean, square the differences, sum them up, and then divide the sum by $n-1$.

The calculation of the sample variance, using Definition 2.2.5, requires two passes through the data. First, we have to calculate the sample mean ($\bar{x}$) and then, once we have that, we must go though the data a second time to calculate the squared differences from the mean for each of the data points. Definition 2.2.6 provides an alternate "shortcut" formula, that is mathematically equivalent, for calculating the sample variance that only requires one pass through the data.

> **Definition 2.2.6 — Sample variance "shortcut" formula.** For a sample of data, $x_1, \ldots, x_n$, a shortcut formula for sample variance, that is mathematically equivalent to the equation in Definition 2.2.5, is
>
> $$s^2 = \frac{1}{n-1} \left[\sum_{i=1}^{n} x_i^2 - \frac{1}{n} \left(\sum_{i=1}^{n} x_i \right)^2 \right].$$

The sample variance is basically the average squared distance of a sample point from the mean. The larger the sample variance, the more the points are spread out around the mean. The smaller the sample variance, the tighter they are around the sample mean.

■ **Example 2.11 — Calculating the sample variance.** Calculate the sample variance for the data from Example 2.1, $\{2.3, 8.1, 5.5, 9.0, 7.8\}$, both manually and in R.
Solution:

$$
\begin{aligned}
s^2 &= \frac{1}{n-1} \sum_{i=1}^{n} (x_i - \bar{x})^2 \\
&= \frac{1}{4} \sum_{i=1}^{5} (x_i - 6.54)^2 \\
&= \frac{1}{4} \left[(2.3 - 6.54)^2 + (8.1 - 6.54)^2 + (5.5 - 6.54)^2 + (9.0 - 6.54)^2 + (7.8 - 6.54)^2 \right] \\
&= \frac{1}{4} [17.9779 + 2.4336 + 1.0816 + 6.0516 + 1.5876] = 7.283
\end{aligned}
$$

Now, in R using the `var()` function:
```
> var(c(2.3,8.1,5.5,9.0,7.8))
[1] 7.283
```
■

The calculation for the population variance, σ^2, is slightly different from the sample variance. As shown in Definition 2.2.7, the population mean replaces the sample mean in the formula and the sum of the squared differences is divided by N, not $n-1$.

> **Definition 2.2.7 — Population variance.** For population data, $x_1, \ldots, x_N$, the population variance is calculated as
>
> $$\sigma^2 = \frac{1}{N} \sum_{i=1}^{N} (x_i - \mu)^2.$$

Note that σ is the Greek letter "sigma" and σ^2 is pronounced sigma-squared.

For both the sample and population calculations, the standard deviation is simply the square root of the variance. Typically the standard deviation is used to characterize variation because it is in the same units as the mean.

> **Definition 2.2.8 — Standard deviation.** The standard deviation the square root of the variance. For a sample it is $s = \sqrt{s^2}$ and for a population it is $\sigma = \sqrt{\sigma^2}$.

As the next example shows, to calculate the sample standard deviation in R, we can use the `sd()` function or take the square root of the results of the variance.

■ **Example 2.12 — Calculating the sample standard deviation in R.** Use R to calculate the sample standard deviation of the data from Example 2.1, $\{2.3, 8.1, 5.5, 9.0, 7.8\}$.

Solution:
```
> sd(c(2.3,8.1,5.5,9.0,7.8))
[1] 2.698703
> sqrt(var(c(2.3,8.1,5.5,9.0,7.8)))
[1] 2.698703                                                                                    ∎
```

As was must mentioned, the standard deviation is most often used to characterize the variation in data because it's in the same units as the data itself. For example, returning to the airline delay data, the standard deviation of *DepDelay* is in minutes, just like the mean and the data, and thus it is interpretable. The variance, on the other hand, is in units of minutes-squared for this data.

Example 2.13 shows the standard deviations in airline delay for 1988, 1997, and 2007, where we see that the variation in the data is increasing from 1988 to 2007 just as the average delay is increasing (see Example 2.2).

■ **Example 2.13 — Using R to calculating the standard deviation of airline delay in 1988.**
```
> sd(data88$DepDelay, na.rm=TRUE)
[1] 21.77714
> sd(data97$DepDelay, na.rm=TRUE)
[1] 28.47112
> sd(data07$DepDelay, na.rm=TRUE)
[1] 36.14189                                                                                    ∎
```

The *range*, R, is another measure of variation.[1] The range has the advantage that it is easy to calculate because it is just the difference between the largest and smallest observations. Using the order statistic notation, the range is defined below.

Definition 2.2.9 — Range.
For a sample of data, the range R is defined as $R = x_{(n)} - x_{(1)}$; for a population it is $R = x_{(N)} - x_{(1)}$.

■ **Example 2.14 — Calculating the range.** Again returning to the data from Example 2.1, {2.3, 8.1, 5.5, 9.0, 7.8}, calculate the range manually and in R.

Solution:
$x_{(1)} = 2.3$ and $x_{(n)} = x_{(5)} = 9.0$, so $R = 9.0 - 2.3 = 6.7$.

Now, in R using the `min()` and `max()` functions:
```
> max(c(2.3,8.1,5.5,9.0,7.8)) - min(c(2.3,8.1,5.5,9.0,7.8))
[1] 6.7                                                                                          ∎
```

In Example 2.14, we used the `min()` and `max()` functions to calculate the range. R also has a `range()` function, but it returns the minimum and maximum values for a set of data, not R. However, as Example 2.15 shows, we can use the `range()` function to calculate R a couple of ways.

■ **Example 2.15 — Calculating the range in R using the `range()` function.** Use the `range()` function to calculate R for the data from Example 2.1: {2.3, 8.1, 5.5, 9.0, 7.8}.

First, here's what the output from the range function looks like:
```
> range(c(2.3,8.1,5.5,9.0,7.8))
```

---

[1]Notation confusion alert: The range is denoted by an italicized capital Roman letter *R*. Don't confuse this with the non-italicized letter R which refers to the R software program.

```
[1] 2.3 9.0
```

Now, using the bracket notation to extract the parts of the `range()` function output, we can calculate *R* via the expression:

```
> range(c(2.3,8.1,5.5,9.0,7.8))[2] - range(c(2.3,8.1,5.5,9.0,7.8))[1]
[1] 6.7
```

Alternatively, we can use the `diff()` function to do the calculation compactly:

```
> diff(range(c(2.3,8.1,5.5,9.0,7.8)))
[1] 6.7
```
                                                                                                              ∎

> **Exercise 2.3** Both manually and in R, calculate the variance, standard deviation, and range of the sample data from Exercise 2.2: {98.6, 92.4, 97.4, 118.6, 84.1, 111.1, 84.1, 103.8, 90.5, 102.4, 101.0, 100.6, 109.6, 102.5, 106.0, 98.7, 79.5, 85.0, 114.1, 89.3}.
>
> *Answer*: $s^2 = 115.821$, $s = 10.762$, and $R = 39.1$.   ∎

2.2.3 Measures of How Two Variables Co-vary

When looking at a data set with more than one variable, it is often natural to ask whether they seem to be related in some way. For two variables *x* and *y*, one such measure is the *sample covariance*.

> **Definition 2.2.10 — Sample covariance.** For a sample of data $\{(x_1,y_1),(x_2,y_2),\ldots,(x_n,y_n)\}$, the covariance is calculated as
>
> $$\text{cov} = \frac{1}{n-1}\sum_{i=1}^{n}(x_i - \bar{x})(y_i - \bar{y}).$$

As Definition 2.2.10 shows, covariance is a measure of how two variables co-vary about their means. Note that the data come in pairs, a pair consisting of one *x* observation and one *y* observation. Each pair is indexed from 1 to *n*. So, x_1 goes with y_1, x_2 goes with y_2, on up to x_n with y_n. The equation, then, essentially calculates the average of the product of the differences between each *x* in the pair from the mean of the *x*s and each *y* in the pair from the mean of the *y*s.

Covariance is a measure of both the strength and direction of the <u>linear</u> relationship between *x* and *y*. If the covariance is a large number (either positive or negative) then the strength of linear association is large, if the covariance is near zero, then the strength of linear association is weak or nonexistent. Similarly, if the sign of the covariance is positive, then the association is positive (meaning the *x* values tend to vary in the same direction as the *y* values); if the sign is negative then the association is negative (meaning the *x* values tend to vary in the opposite direction as the *y* values).

However, it is difficult to know when to call a covariance "large" because it depends on how the observation is measured. That is, changing the measurement units changes the value of the computed covariance. Example 2.16 illustrates this with the airline delay, where we see that the covariance between departure and arrival delays is different when we measure it in minutes or hours. This is troublesome, since changing the measurement units (inches instead of feet; grams instead of kilograms) does not change the association between the *x* and *y*. And, if the association is the same, the numerical measure describing that association should be the same.

∎ **Example 2.16 — Covariance of departure and arrival delays.** Let's first calculate the covariance using the 2007 data where delay is measured in minutes. The R function is cov(), and because there are missing values, we need to use the use="complete.obs" option which tells R to only use those observations where both *DepDelay* and *ArrDelay* are non-missing.

```
> cov(data07$DepDelay,data07$ArrDelay,use="complete.obs")
[1] 1320.23
```

Now, let's change the units of delay to hours and recalculate the covariance.
```
> DepDelay.hrs <- data07$DepDelay/60
> ArrDelay.hrs <- data07$ArrDelay/60
> cov(DepDelay.hrs,ArrDelay.hrs,use="complete.obs")
[1] 0.3667305
```
■

The *sample correlation* solves this problem. Sample correlation is the sample covariance divided by the sample standard deviation of the *x* values and the sample standard deviation of the *y* values. Denoted by *r*, the definition of the sample correlation is shown below.

> **Definition 2.2.11 — Sample correlation.** For a sample of data $\{(x_1,y_1),(x_2,y_2),\ldots,(x_n,y_n)\}$, the correlation is calculated as
>
> $$r = \frac{\text{cov}}{s_x \times s_y} = \frac{\sum_1^n (x_i - \bar{x})(y_i - \bar{y})}{\sqrt{\sum_1^n (x_i - \bar{x})^2}\sqrt{\sum_1^n (y_i - \bar{y})^2}}$$

Dividing by the standard deviations of *x* and *y* makes the correlation independent of the measurement scale, so that the correlation is always between −1 and 1. That makes interpretation much easier. Example 2.17 revisits the airline arrival and departure delays from Example 2.16 and shows that the correlation is the same whether departure is measured in minutes or hours (or any other time unit for that matter). Figure 2.2 illustrates what various levels of correlation look like.

■ **Example 2.17 — Correlation of departure and arrival delays.** Let's now calculate the correlation of the 2007 arrival and departure delays when delay is measured in minutes. The R function is cor() and, as with the cov() function, because there are missing values we again need to use the use="complete.obs" option.

```
> cor(data07$DepDelay,data07$ArrDelay,use="complete.obs")
[1] 0.9315028
```

And the correlation of arrival and departure delays in units of hours is the same.
```
> cor(DepDelay.hrs,ArrDelay.hrs,use="complete.obs")
[1] 0.9315028
```
■

A correlation near 1 is a strong positive linear association and a correlation near −1 is a strong negative linear association, where a correlation of either +1 or −1 is a perfect linear relationship. For example, not surprisingly in Example 2.17 we see that arrival delays are highly correlated with departure delays, which says that if a plane departs late then it is very likely to arrive late as well. A correlation of zero means that there is no linear association between *x* and *y*.

The words "association" and "linear" are purposely used in the previous explanation. It is possible for two variables to be related in a non-linear fashion yet have zero correlation. So, as the next example demonstrates, observing a correlation of zero does not mean there is no association between two variables, only that there is no *linear* relationship.

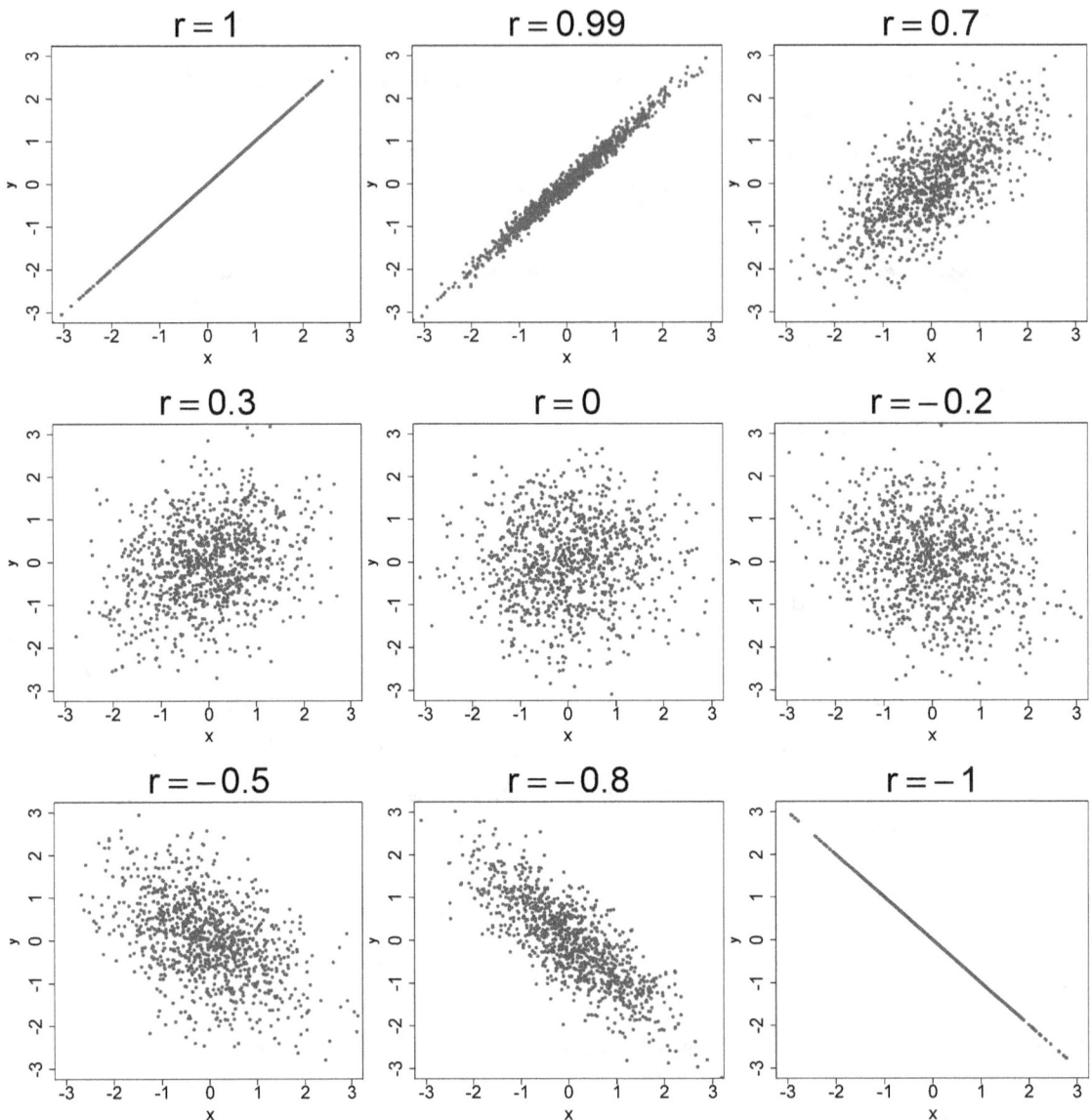

Figure 2.2: *Scatterplots* illustrating various levels of correlation, from $r = 1$ in the upper left, to $r = 0$ in the middle, to $r = -1$ in the lower right. We will learn more about scatterplots in Section 3.2.5 of the next chapter.

■ **Example 2.18 — Causation but zero correlation.** Here's a simple example in which there is clear causation between two variables x and $y - y$ is a direct function of x – and yet the correlation is (essentially) zero.

```
> x <- seq(-1,1,0.1)     # x is a sequence from -1 to 1 in steps of 0.1
> y <- x ∧ 2             # y is a direct function of x
> cor(x,y)               # yet the correlation is virtually zero
[1] 1.216307e-16                                                              ■
```

Now, it's also important to note that a non-zero correlation does not mean there is a causal relationship between the variables. Correlation between the two variables can occur for reasons not associated with direct causality, so in the absence of other information, the most that can be said if a non-zero correlation is observed is that there is an association between the variables. For example, there is a positive correlation between monthly ice cream sales and the number of shark bites in a month. But eating ice cream does not cause an increase in shark bites, nor do shark bites cause an increase in ice cream sales. Rather, both tend to occur more often in the summer and less often in the winter, because people tend to eat more ice cream and they tend to go to the beach when it's hot, which is why there is a correlation.

> **Exercise 2.4** Both manually and in R, calculate the covariance and correlation of the following sample data:
>
> $$x = \{10.9, 10.8, 11.2, 9.1, 11.5, 10.4, 10.5, 10.7, 9.9, 10.1\}$$
> $$y = \{22.4, 23.6, 22.5, 20.6, 23.0, 21.6, 21.6, 21.9, 23.1, 22.2\}$$
>
> *Answer*: $\text{cov} = 0.3561, r = 5912$.

2.2.4 Other Summary Statistics

There are other useful summary statistics in addition to measures of location and variation. One is the *percentile*. The pth percentile is the value of x such that p percent of the data are less than or equal to x. For a given x, the information communicated by its percentile is where x ranks with respect to the rest of the data. That is, it specifies how much of the rest of the data is less than or equal to x.

A standardized testing example will make the idea more concrete. If a person achieves a score of $x = 720$ on the mathematics section of the SAT, the number 720 is not particularly informative unless we have some idea of where it places the person among the rest of those who took the test. Now, if we're told that a score of 720 is the 96th percentile, then it's now clear what the score means: 96 percent of the other test takers scored 720 or less or, conversely, that only four percent of the test takers did better.

The percentile for a particular value of x in a data set can be calculated directly and for other values that that are not observed it can be interpolated. That is, for a particular order statistic $x_{(i)}$ in a sample of data with n observations, the percentile is calculated as described in Definition 2.2.12.

Definition 2.2.12 — Percentiles for sample data. The ith order statistic $x_{(i)}$ is the p_ith percentile of a sample of data with n observations, where

$$p_i = \left(\frac{i-1}{n-1}\right) \times 100.$$

There are some special percentiles. The 100th percentile is the largest (maximum) observation in the data, the median is the 50th percentile, and the 0th percentile is the smallest (minimum) observation. These follow directly from Definition 2.2.12. Also, from this, we can define the range R as the 100th percentile minus the 0th percentile.

■ **Example 2.19 — Calculating the percentile values for sample data.** Find the p_i values for the data from Example 2.1: {2.3, 8.1, 5.5, 9.0, 7.8}.

Solution:

i	p_i	Percentile
1	$(1-1)/(5-1) \times 100 = 0\%$	2.3
2	$(2-1)/(5-1) \times 100 = 25\%$	5.5
3	$(3-1)/(5-1) \times 100 = 50\%$	7.8
4	$(4-1)/(5-1) \times 100 = 75\%$	8.1
5	$(5-1)/(5-1) \times 100 = 100\%$	9.0

So, for example, 5.5 is the 25th percentile, while 8.1 is the 75th percentile.   ■

Note that, as Example 2.19 should make clear, the percentile values assigned to each order statistic are not a function of the actual value of the order statistic. Rather, they are only a function of the sample size (n) and the particular location of the order statistic within all the other data. That is, for $n = 5$, the percentile for the second order statistic will always be 25 percent regardless of the actual value of $x_{(2)}$.

As we previously mentioned, linear interpolation can be used to estimate the percentiles of other values between $x_{(1)}$ and $x_{(n)}$. That is, to calculate the percentile for x, $x_{(1)} < x < x_{(n)}$, find the two order statistics closest to x. Let's call them $x_{(i)}$ and $x_{(i+1)}$, where $x_{(i)} < x < x_{(i+1)}$. Then the percentile for x, let's call it p_x, is

$$p_x = p_i + (p_{i+1} - p_i)\left(\frac{x - x_{(i)}}{x_{(i+1)} - x_{(i)}}\right).$$

Also notice how, with a little algebra, we can rearrange the above expression so that for a given percentile p_x we can find the associated value of x. That is,

$$x = x_{(i)} + (x_{(i+1)} - x_{(i)})\left(\frac{p_x - p_i}{p_{i+1} - p_i}\right).$$

■ **Example 2.20 — Calculating particular percentiles.** For the data from Example 2.1, {2.3, 8.1, 5.5, 9.0, 7.8}, find the 95th percentile.

Solution: Because $p_4 < 95 < p_5$,

$$
\begin{aligned}
x &= x_{(i)} + \left(x_{(i+1)} - x_{(i)}\right)\left(\frac{p_x - p_i}{p_{i+1} - p_i}\right) \\
&= x_{(4)} + \left(x_{(5)} - x_{(4)}\right)\left(\frac{95 - p_4}{p_5 - p_4}\right) \\
&= 8.1 + (9.0 - 8.1)\left(\frac{95 - 75}{100 - 75}\right) \\
&= 8.1 + 0.9 * 0.2/0.25 = 8.82.
\end{aligned}
$$

■

Quantiles are often used in the statistical literature instead of percentiles. Quantiles are simply percentiles divided by 100. That is, while percentiles are on a scale of 0 to 100 (percent), quantiles are on a scale from 0 to 1. Thus, if we denote quantiles by q_i, then $q_i = p_i/100$.

Those quantiles that divide the data into fourths are called *quartiles*, where the first quartile is the same as the 25th percentile or the 0.25 quantile, the second quartile is the median, and the third quartile is the same as the 75th percentile or the 0.75 quantile. The *interquartile range* or IQR is defined as the 75th percentile (equivalently, the 0.75 quantile) minus the 25th percentile (equivalently, the 0.25 quantile).

Appropriately enough, in R, the `quantile()` function is used to calculate quantiles and IQR calculates the IQR. And, remember that a quantile (or percentile) is a value of x. Thus, the quantile function has two arguments: the observed data in the form of a vector and either a single q_i value or a vector of q_i values. The function then returns the quantiles corresponding to the q_i value or values.

■ **Example 2.21 — Using R to calculate quantiles and the IQR.** Using R, confirm the Example 2.20 calculation for the 95th percentile. Then find the quantiles corresponding to the sample data and compare them to Example 2.19. Finally, calculate the IQR.

Solution:
```
> quantile(c(2.3,8.1,5.5,9.0,7.8),0.95)
 95%
8.82
> quantile(c(2.3,8.1,5.5,9.0,7.8),c(0,0.25,0.5,0.75,1))
 0% 25% 50% 75% 100%
2.3 5.5 7.8 8.1 9.0
> IQR(c(2.3,8.1,5.5,9.0,7.8))
[1] 2.6
```
■

Percentiles and quantiles can be quite useful with real data as a way to understand how the data are distributed. In particular, while measures of location tell us what the typical or average value is, and measures of variation tell us something about how spread out the data are around the average value, quantiles and percentiles can provide detailed information about how the data are distributed. Example 2.22 illustrates this idea.

■ **Example 2.22 — Airline delay percentiles.** Returning to the airline delay data, find the 5th and the 95th percentiles of the 1988, 1997, and 2007 data.

Solution:
```
> quantile(data88$DepDelay,c(0.05,0.95),na.rm=TRUE)
```

```
 5%  95%
 -2  35
> quantile(data97$DepDelay,c(0.05,0.95),na.rm=TRUE)
 5%  95%
 -4  48
> quantile(data07$DepDelay,c(0.05,0.95),na.rm=TRUE)
 5%  95%
 -9  74
```
■

What we see in Example 2.22 is that in 1988 five percent of the flights actually left two or more minutes early (where the negative number means that the actual departure time was before the scheduled departure time) and only 5 percent of the flights were delayed for more than 35 minutes. On the other hand, in 1997, the 95th percentile increased to 48 minutes and in 2007 it's up to 74 minutes. So, even though the 5th percentile indicates that some planes are leaving earlier, the delay time for planes at the 95th percentile more than doubled.

An alternative to the first and third quartiles are *hinges*.

> **Definition 2.2.13 — Hinges.** For a sample of size n, the lower and upper hinges are defined as the $x_{(j)}$ and $x_{(n-j+1)}$ order statistics for
>
> $$j = \frac{\left\lfloor \frac{n+1}{2} + 1 \right\rfloor}{2}.$$
>
> If j is not an integer then appropriately interpolate between the two closest order statistics. Here $\lfloor x \rfloor$ is the floor function and it means that if x is not an integer then round x down to the next lowest integer.

An easy way to understand hinges is to see that:
- If n is even, the hinges are the median values of the upper and lower halves of the sorted data.
- If n is odd, the hinges are the median values of the upper and lower halves of the sorted data, *where each half includes the median data point.*

■ **Example 2.23 — Calculating the hinges.** Calculate the lower and upper hinges for the sample data $\{2.3, 8.1, 5.5, 9.0, 7.8\}$ and compare them to the first and third quartiles in Example 2.21.

Solution: For this data, $n = 5$ so

$$j = \frac{\left\lfloor \frac{5+1}{2} + 1 \right\rfloor}{2} = 2.$$

Therefore, the lower hinge is $x_{(2)} = 5.5$ and the upper hinge is $x_{(4)} = 8.1$. These exactly match the first and third quartiles in Example 2.21.
■

■ **Example 2.24 — Airline delay hinges and quartiles.** Returning to the airline delay data, find the hinges and quartiles for the 2007 data.

Solution:
```
> quantile(data07$DepDelay,c(0.25,0.75),na.rm=TRUE)
 25% 75%
 -4  11
> fivenum(data07$DepDelay)
[1] -305 -4 0 11 2601
```
■

In Example 2.23, the R function `fivenum()` returns the minimum, lower hinge, median, upper hinge, and maximum of the data. So, in Examples 2.23 and 2.24 we see that the hinges exactly match the first and third quartiles. This will always be true when n is odd. When n is even they may not match but will typically be quite close.

R also has a `summary()` function that appropriately summarizes whatever object it operates on. We'll have to defer a more complete description of this function until later chapters. However, it is worth noting here that the summary function run on a vector of data returns the minimum, first quartile, median, mean, third quartile, and maximum of the data.

■ **Example 2.25 — Airline delay** `summary()` **statistics.** Returning to the airline delay data, compare the results of the `summary()` function below to the results from Example 2.24.

Solution:

```
> summary(data07$DepDelay)
   Min.  1st Qu.  Median   Mean  3rd Qu.    Max.    NA's
 -305.0    -4.0     0.0   11.4    11.0   2601.0  160748
```

Note that the summary function does not need to be told whether or not to count NA values. It automatically excludes them and provide a count of the number of NA values in the vector. ■

2.3 Longitudinal Data

As we discussed in Section 2.1.1, longitudinal data (also often referred to as time series data) is data that occurs over time. In fact, all data that does not occur at precisely the same moment in time are longitudinal in some sense. But the key distinguishing feature of longitudinal data is that the same subject or subjects are repeatedly measured over time. For example, stock prices are longitudinal data because the price of each of the stocks is recorded at different points in time and, typically, we are interested in how the prices changed over time: are prices rising or falling?

> **Definition 2.3.1 — Longitudinal data.** Longitudinal data are a sequence of observations of the same subject or subjects taken at multiple points in time and which can thus be ordered by time.

In this section we will learn about some basic methods for summarizing longitudinal data. We will then delve more deeply into longitudinal data and how to analyze them in Chapter 19. In that chapter, we will learn how to apply time series modeling to help us understand and characterize how observations change over time.

2.3.1 Statistics for Repeating Cross-sections of Data

The statistics described in Section 2.2 can be applied to longitudinal data just as well as to cross-sectional data. However, when applied to longitudinal data we are summarizing the data in terms of one or more statistics per time period. In fact, this is really what we have been doing up until this point with the airline data when we calculated various statistics for the years 1988, 1997, and 2007. Example 2.26 takes this to its logical conclusion, calculating annual statistics for all 22 years of the airline flight data (although we could go even further and look at sequences of monthly, weekly, or even daily statistics).

■ **Example 2.26** Assess the longitudinal trends in departure delays from 1988 to 2007 by year.

Solution: The table below gives the minimum, first quartile, median, standard deviation, mean, third quartile, and maximum annual values for the departure delay data.

Statistic	1987	1988	1989	1990	1991	1992	1993	1994	1995	1996	
Maximum ($x_{(n)}$)	1439	1439	1646	1439	1439	1439	1439	1439	1439	1438	$\cdots$
3rd Quartile	8	5	8	5	4	5	5	6	7	9	$\cdots$
Mean ($\bar{x}$)	8.1	6.7	8.2	6.9	5.8	5.7	6.1	6.7	8.3	10.0	$\cdots$
Std Deviation (s)	24.0	21.8	23.6	22.5	20.6	20.5	21.4	22.2	25.5	28.8	$\cdots$
Median ($\tilde{x}$)	0	0	0	0	0	0	0	0	0	0	$\cdots$
1st Quartile	0	0	0	-1	-1	-1	-1	-1	-1	-1	$\cdots$
Minimum ($x_{(0)}$)	-1354	-1000	-662	-923	-1084	-661	-675	-109	-150	-125	$\cdots$

	1997	1998	1999	2000	2001	2002	2003	2004	2005	2006	2007	2008
$\cdots$	1618	1800	1740	1435	1692	2119	1582	1882	1930	1752	2601	2467
$\cdots$	7	7	7	10	6	4	2	6	7	10	11	8
$\cdots$	8.2	9.3	9.3	11.3	8.2	5.5	5.3	7.9	8.7	10.1	11.4	10.0
$\cdots$	28.5	30.9	32.3	33.6	28.3	26.1	26.2	26.9	31.2	33.5	36.1	35.3
$\cdots$	0	0	0	0	0	0	0	0	0	0	0	-1
$\cdots$	-2	-2	-2	-2	-3	-4	-4	-4	-4	-4	-4	-4
$\cdots$	-918	-111	-85	-990	-204	-1370	-1410	-1197	-1199	-1200	-305	-534

Looking at the table, over the two decades we see a clear trend in increasing departure delays. This is evident with the mean and the third quartile, though it's not particularly dramatic. With the 25th percentile we also see larger early departures, though again the trend is not particularly dramatic. But these and other aspects of the data are reflected in an increasing standard deviation, which suggests there is more uncertainty in departure times.

Also, note the extreme minimum and maximum values. For example, in the late 1980s and 1990s, many of the maximums are just a minute or two under 24 hours (i.e., 1,440 minutes), which seems suspicious. In the later years, some of the maximum exceeded 24 hours. And some of the early departures are similarly large. What does it mean for a plane to have departed 23-1/2 hours (1410 minutes) early? Clearly some of these data will require further scrutiny.   ■

So, as Example 2.26 shows, one approach to longitudinal data is the repeated calculation of summary statistics for separate subsets of data that correspond to different periods of time, and comparison of such statistics can usefully help discern trends and patterns in the data.

Now, in addition to calculating statistics for repeated cross-sections of data, there are methods specifically created for longitudinal data. Many of these come in the form of statistical models, and we will get to them in Chapter 19. For now, in terms of numerical descriptive statistics, the next two subsections focus on statistics calculated from moving "windows" of data and quantifying correlation over time.

2.3.2 Statistics for Moving Windows of Data

One issue with the repeated cross-sectional statistics approach for summarizing longitudinal data, as in Example 2.26, is that the statistic is only updated at the end of each time period. As a result, all of the detail of what happened within the time period is obscured. One solution to this is to decrease size of the time period (often referred to as a"window") over which the calculations are conducted. That is, in the example, instead of calculating the statistics each year, calculate them every month or perhaps weekly or even daily. However, this can get tedious and, whatever the size of the temporal window, trends within the window will be obscured.

Calculating the statistics using a moving window of the most recent n observations is one way to address this problem. Now, rather than only calculating a statistic once every time period, the statistic is calculated for each new observation using the n most recent observations. For example, the *moving average* $\bar{x}_t$ for observation t is calculated as shown in Definition 2.3.2.

> **Definition 2.3.2 — Moving average.** For a sequence of data, $\{x_1, x_2, \ldots, x_t\}$, for a chosen value of m, the moving average for observation t, $t \geq m$, is
>
> $$\bar{x}_t = \frac{1}{m} \sum_{i=t-m+1}^{t} x_i.$$

If the observations are taken are regular points in time, so that the subscript can be thought of as indexing over time, then $\bar{x}_t$ can be thought of as the moving average at time t.

Definition 2.3.2 assumes that no more than one observation can occur per time period so that all the observations can be sequentially ordered in time. But this might not always be the case. For example, if the times of the observations are not measured precisely enough, or if two or more events can happen at exactly the same time, then it might not be possible to order the observations within particular time periods. In these cases, when there are n_i observations in time period i, the moving average for m time periods is calculated at time t, $t \geq m$, as

$$\bar{x}_t = \frac{1}{\sum_{i=t-m+1}^{t} n_i} \sum_{i=t-m+1}^{t} \sum_{j=1}^{n_i} x_{ij},$$

where x_{ij} is the jth observation in the ith time period. Thus, the above equation is equivalent to Definition 2.3.2, but instead of calculating the moving average in terms of the most recent m observations, it calculates it over all the observations in the past m time periods.

■ **Example 2.27** Here we show the moving averages for the airline delay data for $m = 2, 3, 4,$ and 5 years and compare them to the results of Example 2.26.

Statistic	1987	1988	1989	1990	1991	1992	1993	1994	1995	1996	
$\bar{x}$	8.1	6.7	8.2	6.9	5.8	5.7	6.1	6.7	8.3	10.0	$\cdots$
$\bar{x}_t\ (m=2)$	–	7.0	7.4	7.5	6.3	5.7	5.9	6.4	7.5	9.1	$\cdots$
$\bar{x}_t\ (m=3)$	–	–	7.5	7.3	6.9	6.1	5.9	6.2	7.0	8.3	$\cdots$
$\bar{x}_t\ (m=4)$	–	–	–	7.3	6.9	6.6	6.1	6.1	6.7	7.8	$\cdots$
$\bar{x}_t\ (m=5)$	–	–	–	–	7.0	6.7	6.5	6.2	6.5	7.4	$\cdots$

	1997	1998	1999	2000	2001	2002	2003	2004	2005	2006	2007	2008
$\cdots$	8.2	9.3	9.3	11.3	8.2	5.5	5.3	7.9	8.7	10.1	11.4	10.0
$\cdots$	9.1	8.6	9.2	10.3	9.7	6.9	5.4	6.6	8.3	9.4	10.8	10.7
$\cdots$	8.8	9.1	8.7	9.9	9.6	8.4	6.3	6.3	7.3	8.9	10.1	10.5
$\cdots$	8.3	8.9	9.1	9.5	9.4	8.6	7.5	6.8	7.0	8.0	9.5	10.0
$\cdots$	7.9	8.4	9.0	9.6	9.2	8.7	7.8	7.6	7.2	7.6	8.7	9.6

The table clearly shows that the choice of m affects the moving averages. While it's a bit hard to see in the table, Figure 2.3 shows that the mean ($\bar{x}$) is the most volatile, jumping up and down from one year to the next. However, with increasing m, this volatility is smoothed out. Whether the smoothing is desirable, and what constitutes too much smoothing, is something each analyst must think about carefully. We will learn more about these types of plots in Section 3.3. ■

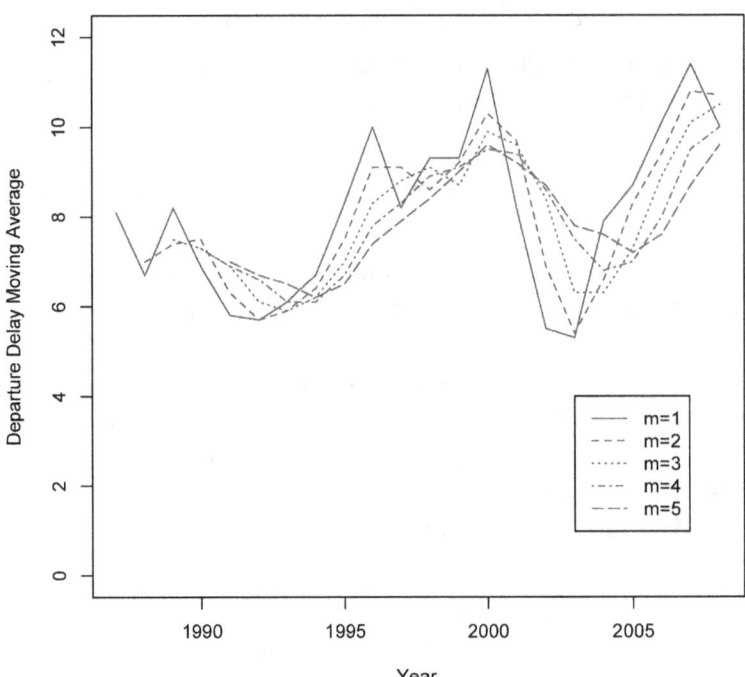

Figure 2.3: A plot of the moving averages from Example 2.27.

The generalization of the idea of calculating a moving average to other statistics should be fairly obvious. For example, assuming there is only one observation per time period, the moving median at time t, $\tilde{x}_t$, is calculated per Definition 2.3.3.

Definition 2.3.3 — Moving median. For a sequence of data, $\{x_1, x_2, \ldots, x_t\}$, for a chosen value of m, the moving median for observation t, $t \geq m$, is

$$\tilde{x}_t = \begin{cases} x_{\left(\frac{2t-m+1}{2}\right)}, & \text{if } n \text{ is odd} \\ \left[x_{\left(\frac{2t-m}{2}\right)} + x_{\left(\frac{2t-m}{2}+1\right)} \right] \big/ 2, & \text{if } n \text{ is even,} \end{cases}$$

The choice of m is critical to the behavior of the moving average, or any other statistic based on a moving window of data. Specifically, the larger m is, the smoother the statistics will be from time period to time period, which may help make trends in the data more visible by smoothing out variation in the data. However, the smoothness comes at the cost of potentially masking short-term changes in the statistic. Thus, with the choice m, we are making an explicit trade-off between the visibility of perhaps subtle long-term trends in the sequence of statistics and the visibility of short-term changes in the statistic.

Exercise 2.5 For $n = 3$, calculate the moving average and moving median of the following sample data:

t	1	2	3	4	5	6	7	8	9	10
x_t	4	1	2	6	6	3	5	5	11	4

Answer:

t	1	2	3	4	5	6	7	8	9	10
$\bar{x}_t$	–	–	2.33	3.00	4.67	5.00	4.67	4.33	7.00	6.67
$\tilde{x}_t$	–	–	2	2	6	6	5	5	5	5

Finally, the choice of a moving window using the past m observations or m time periods is based on the presumption of *prospective analysis*. A prospective analysis is one that only looks forward in time and, in such cases, the only data available is that which is up to the current time period. However, in a *retrospective analysis*, when one is looking back in time, the window can be centered around the time period of interest, using data from both before and after it. For example, assuming m is odd, the moving average calculation in Definition 2.3.2 could be modified for a retrospective analysis as shown in Definition 2.3.4.

> **Definition 2.3.4 — Moving average for retrospective analysis.** For a sequence of retrospective data $\{x_1, x_2, \ldots, x_t\}$, and for some chosen value of $k = 1, 2, 3, \ldots$, the moving average for observation $t \geq m/2 + 1$ is
>
> $$\bar{x}_t = \frac{1}{2k+1} \sum_{i=t-k}^{t+k} x_i.$$

So, the retrospective moving average is averaging over $m = 2k + 1$ observations: x_t plus the k observations immediately before and after it.

Autocorrelation

Autocorrelation is the correlation of a longitudinal data set with itself. What autocorrelation quantifies is the similarity between longitudinal observations as a function of the time separation (or "lag") between them. For some lag k, $k = 1, 2, \ldots$, the sample autocorrelation function r_k is defined below.

> **Definition 2.3.5 — Autocorrelation.** For a sequence of longitudinal data $\{x_1, x_2, \ldots, x_n\}$, and for some integer value of k, $1 \leq k < n$, the autocorrelation for lag k is
>
> $$r_k = \frac{\sum_{i=k+1}^{n} (x_i - \bar{x})(x_{i-k} - \bar{x})}{\sum_{i=1}^{n} (x_i - \bar{x})^2},$$
>
> where $\bar{x}$ is the sample mean taken over the entire sample and typically k is much smaller than n.

Note how r_k in Definition 2.3.5 is very similar to the correlation r in Definition 2.2.11. Here, instead of calculating the correlation between two separate variables x and y, we're calculating the correlation between x and itself k time periods earlier. The main difference in the definitions is in the denominators, where in Definition 2.3.5 the denominator follows because

$$\sum_{i=1}^{n} (x_i - \bar{x})^2 = \sqrt{\sum_{i=1}^{n} (x_i - \bar{x})^2} \sqrt{\sum_{i=1}^{n} (x_i - \bar{x})^2}.$$

Perfect positive or negative autocorrelation at lag k occurs with $r_k = +1$ or $r_k = -1$; $r_k \approx 0$ indicates little to no autocorrelation for lag k. A plot of r_k versus sequential values of k, called a *autocorrelation plot*, helps to show whether there are dependencies present in the data such as long-term linear or other trends, short- and/or long-term cycles, etc. We'll look at how to plot and interpret autocorrelation plots in Section 3.3.3 of Chapter 3.

2.4 Tabular Summaries of Data

For categorical data, sometimes the most useful way to summarize the data is in a table, either in terms of counts, or percentages, or both. Example 2.28 shows a "one-way" table, meaning a table that summarizes one categorical variable.

■ **Example 2.28 — Tabulating airline cancellation codes for 2007.** The table below is a tabulation of the cancellation codes for the 2007 airline data. "A" means the airline cancelled the flight, "B" means the flight was cancelled due to weather, "C" means the flight was cancelled by the National Air System (NAS), and "D" means the flight was cancelled for security reasons. The table shows the number of times flights were cancelled as well as the percentage of cancelled flights for each code.

	\multicolumn{4}{c}{Cancellation Code}	
	A	B	C	D	Total
Count	66,779	61,936	31,995	39	160,749
Percent	41.542	38.530	19.904	0.024	1.0

The table shows that the greatest number of cancellations in 2007 were attributed to the airlines, followed closely by the weather. In comparison, cancellations dues to the NAS were only about half as many as either the airlines or the weather, and the number cancelled for security reasons was a very small fraction of all cancelled flights.   ■

Tables of counts are easy to calculate in R using the `table()` function. For example, the counts in the above table were calculated using the code:

```
> table(data07$CancellationCode[data07$CancellationCode!=""])
```

To interpret the R code, first note that `data07$CancellationCode!=""` produces a logical vector that contains `TRUE` and `FALSE` entries. It is true when the factor label of *data07$CancellationCode* is something other than blank and false otherwise.[2] Enclosing this in the square brackets after the vector *CancellationCode* means R will return all values of that vector with a `TRUE` value in the brackets, meaning in this case it will return all non-blank cancellation codes.

Why bother with this complicated code? Well, if we just ran the code

```
> table(data07$CancellationCode)
```

we would get a table, but the table would contain one more count corresponding to those flights that did not have a cancellation code (i.e., the number of flights that were not cancelled). Try it.

---

[2]Note the subtle difference here between a blank factor label and missing data denoted by `NA`. Key is the concept of a factor, which is how the *CancellationCode* variable is coded, in which character vectors are more compactly stored by translating them into integers, one integer for each unique vector entry, along with "levels" that are the actual character values.

For a variable containing long character strings this can be an efficient way to store the data since each character string is only stored once and integers, which are likely to be much shorter, are used as placeholders in the vector. This storage scheme is transparent if you print the vector to the screen because R automatically substitutes the factor labels for the integer placeholders before printing the results.

So, when R read in the cancellation code data, it created *CancellationCode* as a factor variable. In so doing, all of the fields missing a cancellation code were assigned the integer 1 with corresponding label `""`. The result is that there are no missing values in the vector and, thus, to identify which flights were not cancelled we must use the syntax `data07$CancellationCode!=""`.

This methodology is a very useful and powerful for subsetting data in R. You can write virtually any logical statement in the square brackets and those observations for which the logical statement is true are then selected. For example,

```
> table(data07$CancellationCode[data07$UniqueCarrier=="AA"])
```

would produce a table of cancellation codes for American Airlines flights (for whom the carrier code is "AA").

Exercise 2.6 In R, tabulate the number of 2007 flights delayed for more than three hours by day of the week.

Answer:

	Monday	Tuesday	Wednesday	Thursday	Friday	Saturday	Sunday
Count	23,461	19,400	21,871	27,074	26,235	14,141	22,010

In addition to one-way tables, we can also create two-way or higher tables that give the counts or percentages (or both) for two or more variables. Example 2.29 is a two-way table of cancellation codes by day of the week.

■ **Example 2.29 — A two-way table: airline cancellation codes by day of the week.** The table below shows the number of airlines cancelled by day of the week for each cancellation code.

	Cancellation Code				
Day of the Week	A	B	C	D	Total
Monday	11,314	6,607	4,346	20	22,287
Tuesday	10,612	9,563	5,404	0	25,579
Wednesday	10,196	10,113	6,114	3	26,426
Thursday	9,607	8,891	6,634	14	25,146
Friday	8,923	10,013	5,281	0	24,217
Saturday	7,238	7,402	1,348	2	15,990
Sunday	8,889	9,347	2,868	0	21,104
Total	66,779	61,936	31,995	39	160,749

The table shows, for example, that overall the greatest number of cancellations occurs mid-week and the fewest on the weekend, but for carrier-related cancellations, the most cancellations occurred on Monday. ■

To read the table in Example 2.29, note that each "cell" in the table gives the count for the number of flights that were cancelled for the reason at the top of the column and for the day of the week corresponding to the row of the cell. For example, 10,013 flights were cancelled on Friday in 2007 for weather-related reasons (cancellation code B).

Along the bottom and the right side of the table are the marginal counts for each of the individual categorical variables. So, for example, the counts the right side provide information only about the the number of cancellations on each day of the week without regard to the reason the flight was cancelled. Similarly, the counts along the bottom provide information only about the reason a flight was cancelled without regard to the day of the week on which is was cancelled (and thus these counts match those in the one-way table of Example 2.28).

Creating a two-way table in R just requires adding a second categorical variable into the `table()` function. For example, the table in Example 2.29 (without the marginal totals) was calculated using:

```
> cancellation.codes <- factor(data07$CancellationCode[data07$CancellationCode
  != ""])
> cancellation.dow <- data07$DayOfWeek[data07$CancellationCode != ""]
> table(cancellation.dow,cancellation.codes)
```

In addition to the `table()` function for creating tables of counts, the `prop.table()` function is useful for calculating and tabulating proportions. For example, if we wanted to convert the counts in table in Example 2.29 into row percentages, so we could get better insight into the distribution of cancellation codes for each day of the week, we would run the following:

```
> prop.table(table(cancellation.dow,cancellation.codes),1)
```

On the other hand, if we wanted to understand into the distribution of each type of cancellation code over the days of the week (i.e., we wanted column percentages), we would run the following:

```
> prop.table(table(cancellation.dow,cancellation.codes),2)
```

Run the code to see what it produces. Also, if we already have a table object, then the `margin.table()` function is useful for calculating the various marginal totals.

2.5 Problems

Problem 2.1 In your own words, describe the difference between quantitative and qualitative data and between cross-sectional and longitudinal data. Give an example of each.

Problem 2.2 In your own words, describe the difference between continuous and discrete quantitative data. Give an example of each. Now describe the difference between ordinal and nominal qualitative data. Give an example of each.

Problem 2.3 In your own words, describe the difference between categorical and ordinal data. Give an example of discrete ordinal data, categorical ordinal data, and nominal categorical data.

Problem 2.4 Using the data taxonomy of Figure 2.1, classify what type of data is contained in the following airline dataset variables: *AirTime*, *Origin*, *Cancelled*, *Month*, and *TailNum*.

Problem 2.5 Calculate the mean, median, and 20 percent trimmed mean manually for the following sample data:
$\{8, 10, 1, 7, 12, 25, 6, 12, 2, 47, 9, 12, 5, 11\}$.
Confirm your results using R.

Problem 2.6 Calculate the mean, median, and 10 percent trimmed mean manually for the following sample data:
$\{-1.5, -3.3, -42.7, 2.9, 7.4, 0.9, -1.2, 0.3, 4.5\}$.
Confirm your results using R.

Problem 2.7 "By hand," calculate the variance and standard deviation of the data from Problem 2.5. Confirm your results using R.

Problem 2.8 "By hand," calculate the variance and standard deviation of the data from Problem 2.6. Confirm your results using R.

Problem 2.9 "By hand," calculate the 25th and 75th percentiles and the lower upper hinges of the data from Problem 2.5. Confirm your results using R.

Problem 2.10 "By hand," calculate the 25th and 75th percentiles and the lower upper hinges of the data from Problem 2.6. Confirm your results using R.

Problem 2.11 Calculate the covariance and correlation manually for the following data:
$x = \{-0.779, 0.253, 0.529, 1.632, 1.332, -2.167\}$
$y = \{-0.415, 1.482, 2.181, 4.240, 3.633, -3.332\}$.
Confirm your results using R.

Problem 2.12 Calculate the covariance and correlation manually for the following data:
$x = \{6.1, 2.3, 1.6, 5.1, 4.4, 7.8, 5.5\}$
$y = \{-2.5, -2.8, 0.5, -2.9, -1.7, -2.5, -3.8\}$.
Confirm your results using R.

Problem 2.13 Calculate the covariance and correlation manually for the following data:
$x = \{1.0, 0.04, 0.16, 0.36, 0.64, 0.36, 0.04, 0.0, 0.16, 0.64, 1.0\}$
$y = \{-1.0, 0.2, 0.4, 0.6, 0.8, -0.6, -0.2, 0.0, -0.4, -0.8, 1.0\}$.
Confirm your results using R.

Problem 2.14 Using Definition 2.3.2, calculate the moving average with $n = 5$ for the following sequence of data:
$\{13, 26, 11, 20, 21, 25, 15, 13, 21, 24, 31, 21, 22, 14,$
$22, 22, 30, 17, 23, 16, 20, 32, 21, 31, 24, 24, 20, 18,$
$19, 17, 15, 15, 15, 31, 25, 20, 13, 18, 11, 18, 10, 12\}$

Problem 2.15 Using Definition 2.3.3 and the data from Problem 2.14, calculate the moving median with $n = 5$.

Problem 2.16 Using Definition 2.3.2, calculate the moving average with $n = 3$ for the following sequence of data:
$\{0.74, 0.91, 0.10, -0.46, -1.13, -1.17, -0.03, -0.67,$
$-0.04, -0.58, 1.09, 0.72, 0.26, 0.49, -0.33, -0.60,$
$-0.62, -0.60, -0.01, 0.08, 0.90\}$

Problem 2.17 Using Definition 2.3.3 and the data from Problem 2.16, calculate the moving median with $n = 4$.

Problem 2.18 Using R, calculate the 2007 aircraft flight arrival delay (the *ArrDelay* variable) mean, median, and 10 percent trimmed mean. Also calculate the variance, standard deviation, and range for the 2007 flight times.

Problem 2.19 Using R, calculate the 2007 aircraft flight time (the *AirTime* variable) mean, median, and 10 percent trimmed mean. Also calculate the variance, standard deviation, and range for the 2007 flight times.

Problem 2.20 Using R, calculate the covariance and correlation between the 2007 aircraft flight arrival delay (the *ArrDelay* variable) and the aircraft flight time (the *AirTime* variable). Are the two variables highly correlated?

Problem 2.21 Using R, calculate the covariance and correlation between the 2007 aircraft flight time (the *AirTime* variable) and the distance flown (the *Distance* variable). Are the two variables highly correlated?

Problem 2.22 Using R, for the 2007 aircraft data construct a two-way table of counts of *Cancellation-Code* versus *Diverted*. What do you learn from the table?

Problem 2.23 Using R, for the 2007 aircraft data construct a two-way table of percentages of *Diverted* versus *Month*. What do you learn from the table?

Problem 2.24 A number of data sets come with the base package in R. (Type data() to view them.) One is called women and it contains the heights and weights of 15 women. (Type women to view the data and ?women to get some information about the data.) Describe and summarize the data using appropriate numerical descriptive statistics.

Problem 2.25 The faithful data set installed with the R base package has data on 272 eruptions of Old Faithful. The data set contains the length of each eruption (in minutes) and the time between eruptions, also in minutes. (Type faithful to view the data and ?faithful for more information about the data.) Describe and summarize the data using appropriate numerical descriptive statistics.

Problem 2.26 The cars data set installed with the R base package has data on 50 observations of car speed (in miles per hour) and distance to stop (in feet). Type cars to view the data and ?cars for more information about the data. Describe and summarize the data using appropriate numerical descriptive statistics. What is the correlation between speed and distance? Do you notice anything unusual about the data?

Problem 2.27 The mtcars data set installed with the R base package has data on 32 cars from the early 1970s. Type cars to view the data and ?cars for more information about the data. Describe and summarize each of the variables in the data using appropriate numerical descriptive statistics. Which of the other 10 variables is most positively correlated with miles per gallon (*mpg*)? Most negatively correlated? (Hint: The cor() function run on the whole data frame will produce a "correlation matrix" that summarizes all pairwise correlations.) Are any of the cars unusual with respect to the other cars in the data set? How so?

Problem 2.28 Prove that the shortcut formula for the sample variance in Definition 2.2.6 is mathematically equivalent to the formula in Definition 2.2.5.

Problem 2.29 Clearly $s^2 \geq 0$. Prove that $s^2 = 0$ if and only if $x_i = \bar{x}$ for $i = 1, 2, \ldots, n$.

Problem 2.30 Some ask why in the sample variance formula the differences are squared. Prove that $\frac{1}{n-1} \sum_{i=1}^{n} (x_i - \bar{x}) = 0$ for any $x_1, x_2, \ldots, x_n$.

Problem 2.31 Clearly $\sigma^2 \geq 0$. Prove that $\sigma^2 = 0$ if and only if $x_i = \mu$ for $i = 1, 2, \ldots, N$.

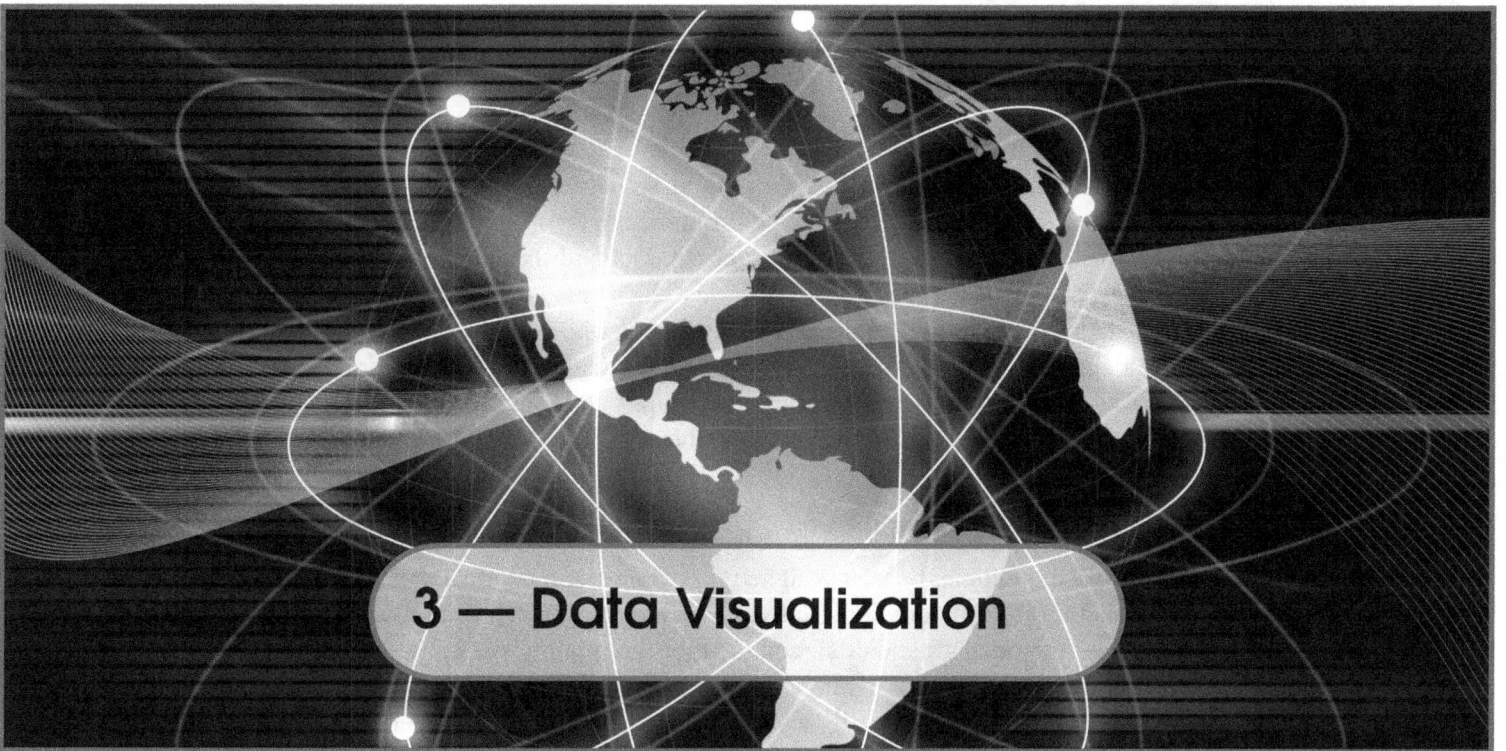

3 — Data Visualization

3.1 Introduction

Graphical plots are the means by which data are most easily visualized and thus understood. Indeed, there is no better tool for finding patterns in data than the human eye applied to appropriate displays of relevant data, particularly patterns that are ill specified or unknown.

Done correctly, good graphs can facilitate:

- *Insight*: Plotting data can enhance perception and comprehension of phenomena, particularly relationships among variables and trends over time, that may not be apparent otherwise. With large data sets or complicated data (or both), appropriate visualizations can make the data easier to understand.
- *Exploration*: With the appropriate analytical software (such as R) that facilitates interaction with data, the data can examined and evaluated in a variety of ways. For subject matter experts and data scientists, interactive methods allow them to bring their expertise to bear in the visualization of the data.
- *Impact*: Data visualization is a powerful way to communicate information, relationships, and even the stories present in data. It's the the old adage that a picture is worth a thousand words applied to data: a good graph is worth thousands of data points.

Good graphs of data can reveal relationships in the data that are simply not apparent via summary statistics. The canonical example is Anscombe's data plotted in Figure 3.1. (These particular plots are called scatterplots; we'll learn more about them in Section 3.2.5.) Visually these four sets of data clearly show very different relationships between the x and y variables, yet:

- the means and standard deviations of each of the x variables are exactly the same;
- the means and standard deviations of the y variables are also the same;
- the correlations between x and y in each of the four cases are the same; and,
- even the regression fits are the same.[1]

---

[1] You will learn about regression modeling in Chapters 17 and 18.

In the absence of plotting the data, by just looking at some descriptive statistics one could be completely misled into thinking that there is little difference in the underlying phenomena.

In summary, there are clear benefits from graphing data and these benefits increase in direct relation to the size of the data though, as we've just seen, even with small data sets such as Anscombe's visualization can provide unique benefits. But unlike Anscombe's data, which is small and simple to visualize via scatterplots, data scientists are frequently faced with large, complicated, and messy data sets. In this situation, good statistical graphs and other forms of data visualization are critical to understanding the data and underlying phenomena.

3.2 Traditional Statistical Graphics

Statistical graphics have traditionally been designed for displaying numerical data. Common to most of these methods is that the data are quantitative, typically with a small number of observations and variables (at least by today's standards), and the plots are mainly static. We'll branch out from traditional graphics in Section 3.4, but it's important to recognize that the graphics we discuss in this section are still in widespread use, are still very useful, and they are quite powerful.

The graphical methods presented in this section are useful compliments to the cross-sectional numerical descriptive statistics in Section 2.2, particularly for further summarizing and displaying such cross-sectional data. As Anscombe's example illustrates in Figure 3.1, when appropriately designed, implemented, and presented, these graphical methods give the user intuitive insights into the data that may not be achievable otherwise.

3.2.1 Bar Charts

Bar charts are useful for summarizing categorical data, particularly for visually comparing the relative sizes of the various categories. Bar charts typically display the category titles on one axis and either counts or percentages for each of the categories on the other.

For example, Figure 3.2 plots the percentage of 2007 cancelled flights by cancellation code. This is a graphical representation of the percentages first tabulated in Example 2.28. As before, the "A" code means the airline cancelled the flight, "B" means the flight was cancelled due to weather, "C" means the flight was cancelled by the National Air System (NAS), and "D" means the flight was cancelled for security reasons. In the plot, we clearly see that flights were cancelled by the carriers more often than any other reason, but it is followed closely by weather. We can also see that the number of NAS-related cancellations are roughly half as big as either carrier- or weather-related cancelations and that the number of security-related cancellations is very small.

Rather than plotting percentages, bar charts can also show the counts per category. For example, Figure 3.3 shows plots equivalent to Figure 3.2 with counts. Note how we can plot the bars either vertically or horizontally. When the category names are long, it is often preferable to plot the bars horizontally so that the names can also be written out horizontally on the y-axis.

Also note that it is virtually impossible to discern the precise number of flights cancelled in each category in Figure 3.3, but that's not the point of the bar chart. Instead, the goal is to give some insight into the overall distribution of data by cancellation code and for that the bar chart works quite well. If it's the precise counts we want, then a table such as Table 2.28 would be a better way to summarize the data.

As shown in Figure 3.4, bar charts can also display subgroupings, either by breaking the bars up to show the constituent subgroups in a stacked bar chart (on the left), or as a set of bars in a side-by-side bar chart (on the right). Stacked bar charts facilitate comparing between the main

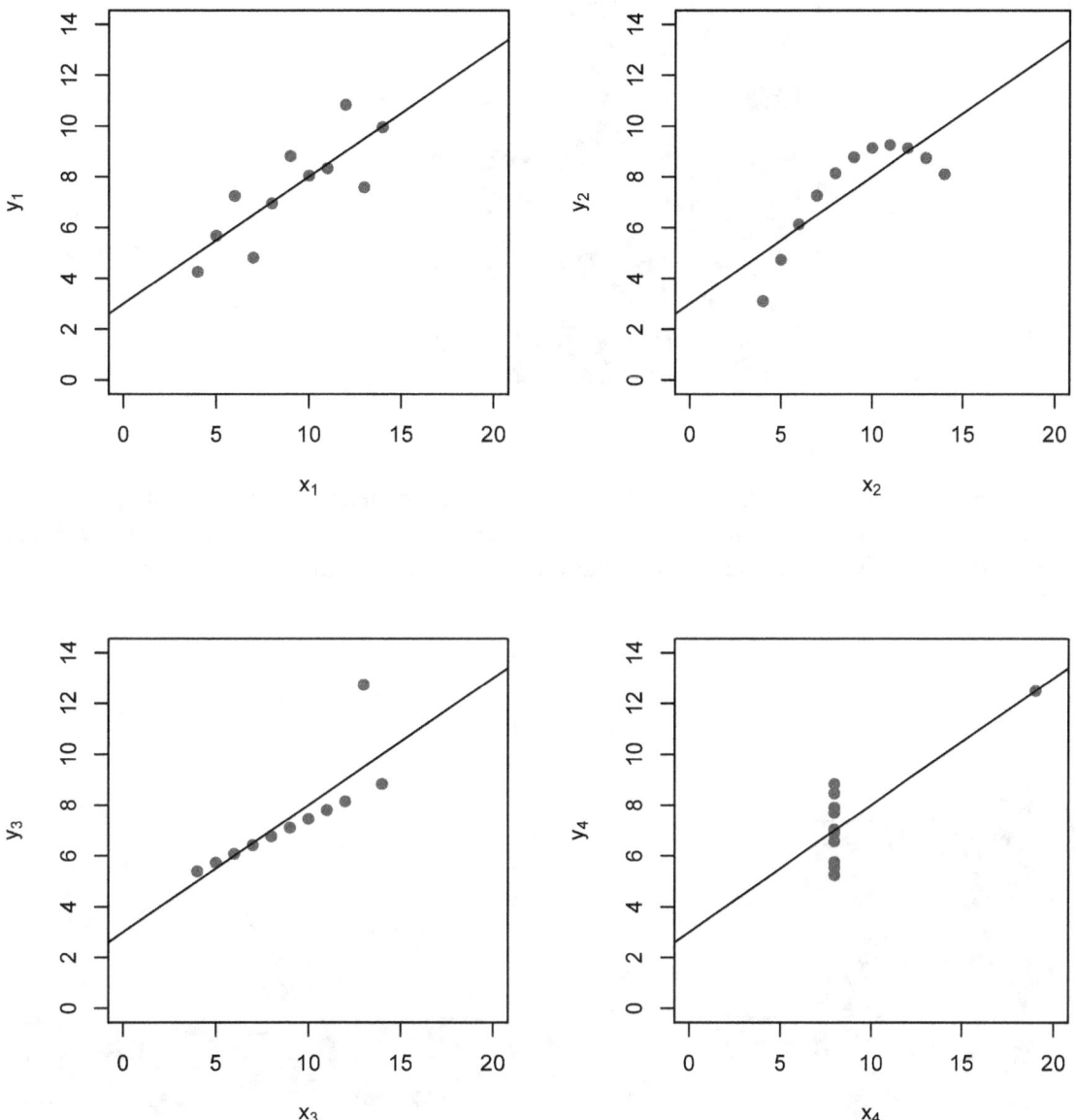

Figure 3.1: Plots of Anscombe's data, where all the descriptive statistics match, the regression fits (as shown by the lines) are the same, and yet the relationships between the *x* and *y* variables are clearly different in all four cases.

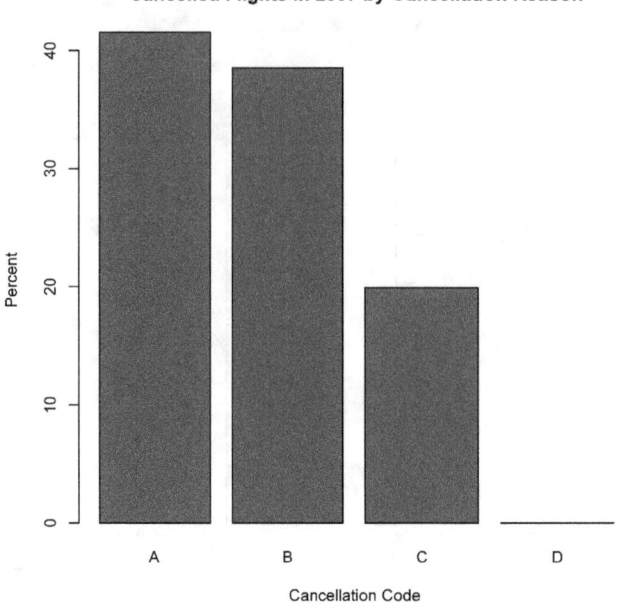

Figure 3.2: Bar chart example: Percent of flights that were cancelled in 2007 by flight cancellation code. "A" means the airline cancelled the flight, "B" means the flight was cancelled due to weather, "C" means the flight was cancelled by the National Air System, and "D" means the flight was cancelled for security reasons.

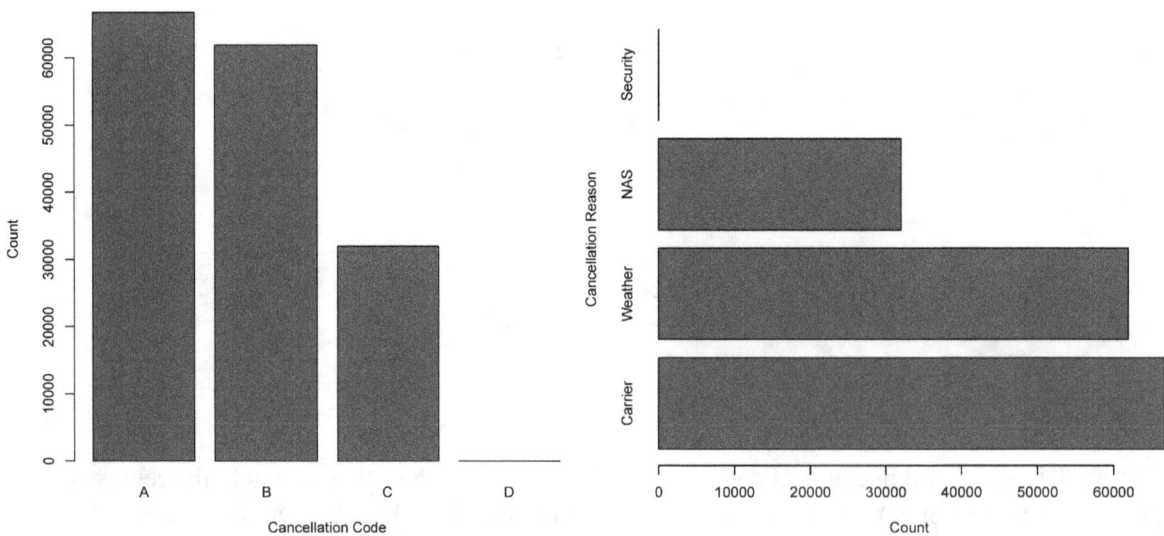

Figure 3.3: Examples of bar charts: Number of flights that were cancelled in 2007 by flight cancellation code. See Table 2.28 for the exact counts.

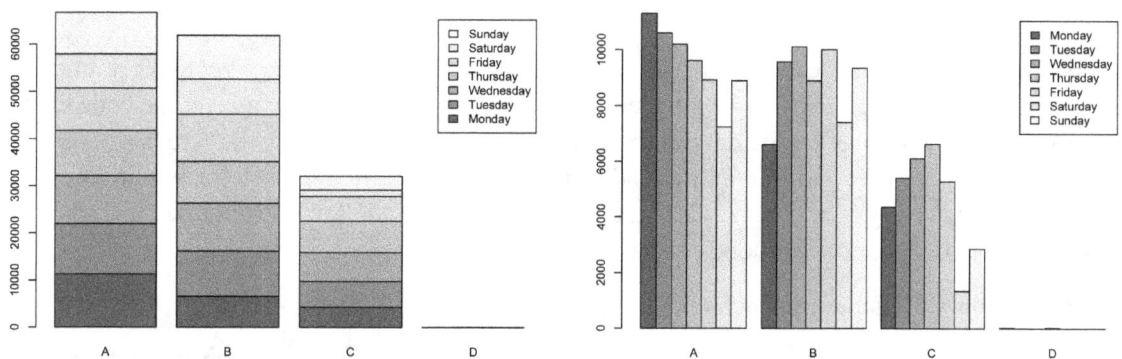

Figure 3.4: Examples of stacked and side-by-side bar charts: Number of flights that were cancelled in 2007 by flight cancellation code and day of the week. The left plot is a stacked bar chart and the right is a side-by-side bar chart. See Table 2.29 for the exact counts.

groupings (cancellation code in the figures) while also allowing for comparison of the relative sizes of the subgroupings within each main group (day of the week in the figures). In contrast, side-by-side bar charts allow direct comparison of the sizes of the subgroupings within the main groupings, but this comes at the cost of not being able to directly compare the sizes of the main groups. Note how, for example, we can easily see that NAS-related delays are smallest on Saturdays and then steadily increase as the week goes on. In contrast, carrier-related cancellations are highest on Monday and then steadily decrease until Saturday and then have a small increase on Sunday.

In R, bar charts are created using the `barplot()` function. Key to using the function is that it takes as inputs the aggregate counts or percentages for each category. Thus, you must typically calculate the counts or percentages ahead of time, typically using the `table()` function. Example 3.1 provides some of the code used to generate Figures 3.2 through 3.4.

■ **Example 3.1 — Bar Chart R Code.** The easiest bar charts to plot are those based on counts. The code below generates the left plot in Figure 3.3.

```
> cancellation.counts <- table(data07$CancellationCode)[2:5]
> barplot(cancellation.counts,xlab="Cancellation Code",ylab="Count",col="blue")
```

The creation of the `cancellation.counts` vector is not really necessary but is used just to make the example a little more readable. The use of `[2:5]` after the table command tells R to only assign the second through fifth entries of the table to `cancellation.counts`. The first entry in the table is the count of all flights without cancellation codes – i.e., the flights that were not cancelled – which are not relevant to this particular plot.

The code below generates the right plot in Figure 3.3. Here we use the `horiz` option to plot horizontal bars and the `names.arg` option to replace the cancellation codes with the cancellation categories.

```
> barplot(cancellation.counts,xlab="Cancellation Code",ylab="Count",
    horiz=TRUE,names.arg=c("Carrier", "Weather", "NAS", "Security"),col="blue")
```

Generating the bar charts in Figure 3.4 is a bit more complicated. First, let's extract the data that we will plot:

```
> cancellation.codes <- factor(data07$CancellationCode[data07$CancellationCode!=""])
> cancellation.dow <- data07$DayOfWeek[data07$CancellationCode!=""]
```

Again, the creation of the `cancellation.codes` and `cancellation.dow` vectors are not really necessary but they make the example a little more readable. In the first line above, the code extracts all the values in the *CancellationCode* vector that are not blank (i.e., for which there is a code entry). In a similar way, the second line extracts all the values in the *DayOfWeek* for which associated *CancellationCode* entry is not blank. The result is two vectors of the same length, where each pair of entries in the two vectors corresponds to the same flight.

The code below then generates each of the plots in Figure 3.4:

```
> barplot(table(cancellation.dow,cancellation.codes),legend.text=c("Monday",
  "Tuesday","Wednesday","Thursday","Friday","Saturday","Sunday"),col="blue")
> barplot(table(cancellation.dow,cancellation.codes),legend.text=c("Monday",
  "Tuesday","Wednesday","Thursday","Friday","Saturday","Sunday"),col="blue",)
  beside=TRUE)
```

Key to generating the stacked and side-by-side bar charts is that the first argument in the `barplot()` function is a two-way table. The default is then a stacked bar chart and a side-by-side plot results by setting the `beside` option to `TRUE`. Also, the legends in the plot are created using the `legend.text` option.   ∎

Pie Charts

Pie charts are a very commonly used graphic for displaying the the relative sizes of categories within a set of data. Figure 3.5 is an example, where this particular plot is an alternative to Figure 3.2. Statisticians prefer bar charts to pie charts because research has shown that people are able to more accurately extract information from bar charts than pie charts. In particular, people are better able to compare the relative sizes of the categories with bar charts. What it comes down to is that human beings are better at accurately comparing lengths than areas and angles. For example, in Figure 3.2 it is clear that a larger percentage of cancellations are attributable to carriers than weather, but the difference is not discernible in the pie chart of Figure 3.5.

∎ **Example 3.2 — Pie Chart R Code.** Using the *cancellation.counts* vector created in Example 3.1, the code below generates the pie chart in Figure 3.5.

```
> pie(cancellation.counts,col=c("light gray","gray","dark gray","black"))
```

The `col` option is used to recolor the pie slices. The default colors are shades of pastel.   ∎

> Exercise 3.1 Using the 2007 airline data, create a bar chart and pie chart showing the number of flights by month. Which plot makes it easier to identify the month with the greatest number of flights? The month with the fewest number of flights?   ∎

3.2.2 Histograms

A histogram is akin to a bar chart but for continuous data. As just discussed, bar charts are for discrete data, and hence each bar in the chart corresponds to a distinct category in the data. In contrast, the histogram is applied to continuous data by dividing the real line up into contiguous ranges (typically called "bins") for which the number of observations that fall into each bin are

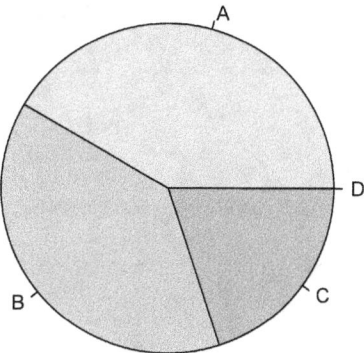

Figure 3.5: Example pie chart: Distribution of flights that were cancelled in 2007 by flight cancellation code. Compare this to the bar chart in Figure 3.2. The R code to generate this figure is shown in Example 3.2.

summed up. Then bars are plotted, where the height of each bar corresponds the number (or fraction) of observed values that fall in the range of the bar.

Note the visual difference between bar charts and histograms. With bar charts, the bars are physically separated from each other. That is, there is space between the bars. However, with histograms the bars are contiguous – they touch. This is because the bins upon which the bars are based are contiguous.

Histograms are often used to gain insight into the distribution of the data. What is its shape? Where are most of the observations located? How spread out is the data? Are there any unusual outliers or concentrations of the data?

Figure 3.6 is a histogram of the distance of 2007 flights showing the number of flights within various mileage bands. The numerical summary statistics for this data are: $\bar{x} = 719.8$ and $\tilde{x} = 569.0$ miles with $s = 562.3$ and $R = 4,951$ miles. These statistics are roughly visually evident in the histogram, which shows a bump around the median, with quite a bit of variation around it, and extreme values that extend from zero to almost 5,000 miles.

In Figure 3.6, the heights of the bars denote the number of flights in each mileage range. Visually prominent in the plot is a spike at the 200-400 mileage range, as well as that all of the higher bars are in the 0-1,000 mile range, indicating that most flights in 2007 were 1,000 miles or less. The skew of the distribution is to the right, meaning that the right "tail" of the distribution is much longer than the left "tail" (which is bounded by zero – distance must be non-negative). This skew is both visually evident in the plot and in the summary statistics with the mean greater than the median ($\bar{x} > \tilde{x}$). Had the distribution been skewed to the left then the mean would have been less than the median.

■ **Example 3.3 — Histogram R Code.** Using the *Distance* variable in the 2007 data, the code below generates the histogram in Figure 3.6.

```
> hist(data07$Distance,xlab="Distance (miles)",ylab="Counts",
   main="Distribution of 2007 Flight Distances",col="royal blue")
```

Note that the xlab and ylab options are used to label the axes, the main option adds the title at the top, and the col option is used to color the histogram bars. The default color is white.   ■

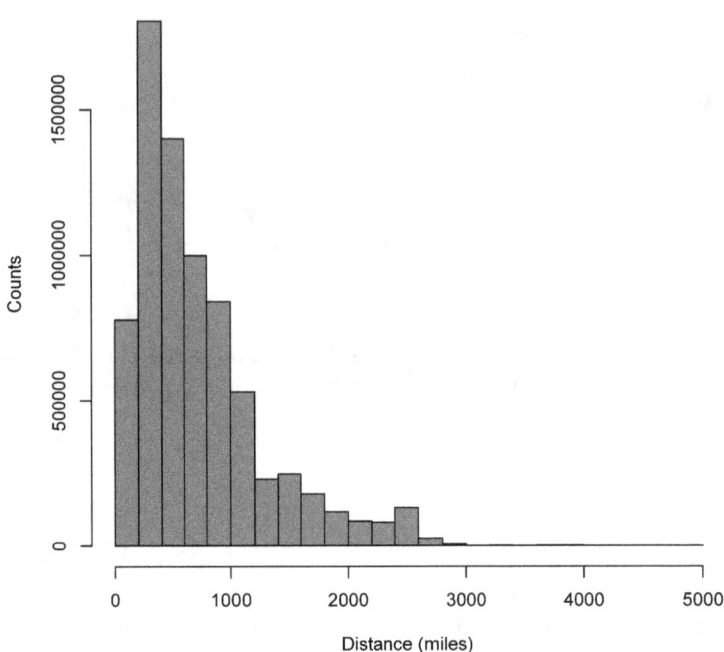

Distribution of 2007 Flight Distances

Figure 3.6: Example histogram: Distribution of flight distances for all 2007 flights.

Given the ubiquity of statistical software, we will ignore the details of how to manually construct a histogram. However, be aware that every software package makes certain choices when drawing a histogram based on default values in the software. Two important choices that need to be made when drawing a histogram are: (1) how many "bins" should be used, and (2) where those bins should be located. A common choice is to begin the first bin at the data's minimum value and to end the last bin at the maximum value. In terms of the right number of bins, when n is relatively small a good choice is often $\sqrt{n}$, and when n is large a good choice is often $10\log_{10} n$.

Regardless of the defaults, it is important to try out alternatives, particularly in terms of the number of bins, since the information communicated with the plot can vary with these choices. For example, Figure 3.7 shows three different histograms for the distribution of 2007 flights originating from the Monterey, California airport (airport code "MRY"). The difference is in the number of bins, where the left histogram has 8 bins, the middle has 15 bins, and the right has 30 bins (set using the breaks option of the hist() function). What the figure shows is that the shape of the distribution, and thus the visual information communicated, depends on the choices made when drawing the histogram.

■ **Example 3.4 — R Code for the Histograms in Figure 3.7.** Using the *AirTime* variable in the 2007 data, the code below generates the histograms in Figure 3.7.

```
> hist(data07$AirTime[data07$Origin=="MRY"],xlab="Flight Time (minutes)",
    ylab="Counts",main="",col="royal blue",breaks=8)
> hist(data07$AirTime[data07$Origin=="MRY"],xlab="Flight Time (minutes)",
```

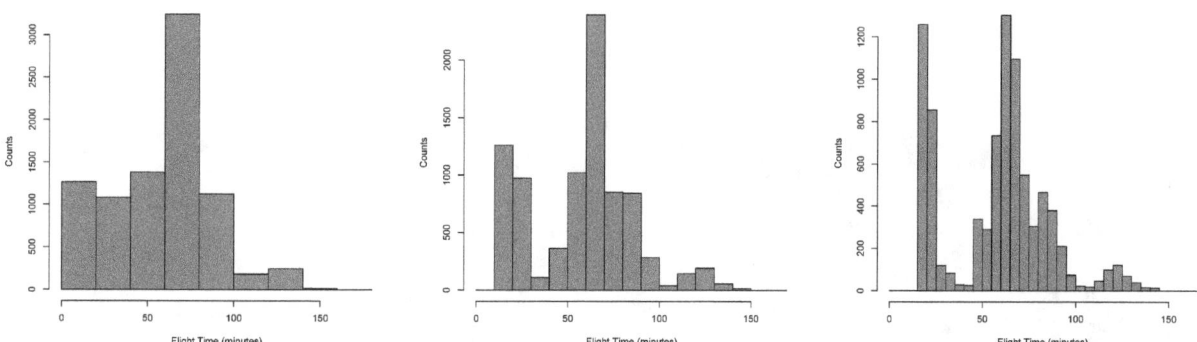

Figure 3.7: Histograms with varying numbers of "bins" showing the distribution of flight times for 2007 flights originating from the Monterey, California airport. The left histogram has 8 bins, the middle has 15 bins, and the right has 30 bins.

```
   ylab="Counts",main="",col="royal blue",breaks=15)
 > hist(data07$AirTime[data07$Origin=="MRY"],xlab="Flight Time (minutes)",
   ylab="Counts",main="",col="royal blue",breaks=30)
```

The `breaks` option allows user input to the number of histogram bins, but the input is only a suggestion and R may override it.   ∎

Now, while it's important to vary the number of bins in order to find the histogram that most appropriately communicates what is important in a given set of data, it's also important to be careful of extremes. Figure 3.8 shows the most extreme histograms for the same flight time data from Figure 3.7. The left plot is a trivial "histogram" with just one bar that shows the total number of flights for all flight times. This "histogram" says nothing about the underlying distribution of flight times and can be more efficiently summed up numerically: there were 8,554 flights in 2007 that originated in Monterey with flight times from 4 to 161 minutes. The right "histogram" is really just a bar chart with a discrete bar for every possible flight time.

Sometimes, because of outliers, it can be useful to plot the data rescaled. For example, the left histogram in Figure 3.9 plots the 2007 delay times for all flights with delay times greater than zero. Because of a small number of very large delay times, virtually off of the data are compressed into the one bar on the far left. This doesn't provide much insight into the distribution of most of the data. Now, we can rescale the plot's *x*-axis to "zoom in" on the smaller times as shown in the middle histogram in Figure 3.9. This is better, but we're left some of the data out of the display. However, another alternative is to transform the data and plot the transformed values. For example, the right histogram in Figure 3.9 shows the log (base 10) transformation of delay times. The log transformation has the effect of compressing the larger delay times and spreading out the smaller delay times. Here we see that plotting data in units other than the natural units can be useful in picturing all the data. In such cases, transformations can reveal interesting aspects of the data.

Finally, note that the R `hist()` function defaults to *frequency histograms* that show counts (i.e., frequencies) on the *y*-axis. That's what we have been looking at in Figures 3.6 through 3.9. Setting the `freq` option to `FALSE` tells R to create a *density histogram* where the areas of all the bars add up to 1. This type of histogram will become important when we start learning about probability distributions in later chapters.

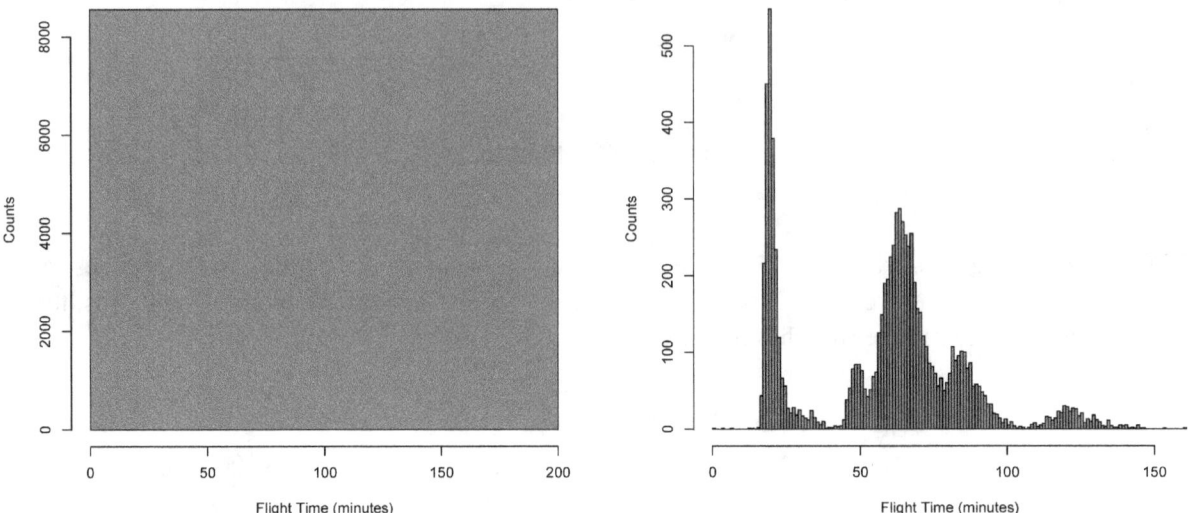

Figure 3.8: Extreme histograms for the same flight time data from Figure 3.7. The left plot is a trivial "histogram" with just one bar that shows the total number of flights for all flight times. The right plot is basically just a bar chart since there is a discrete bar for each and every minute.

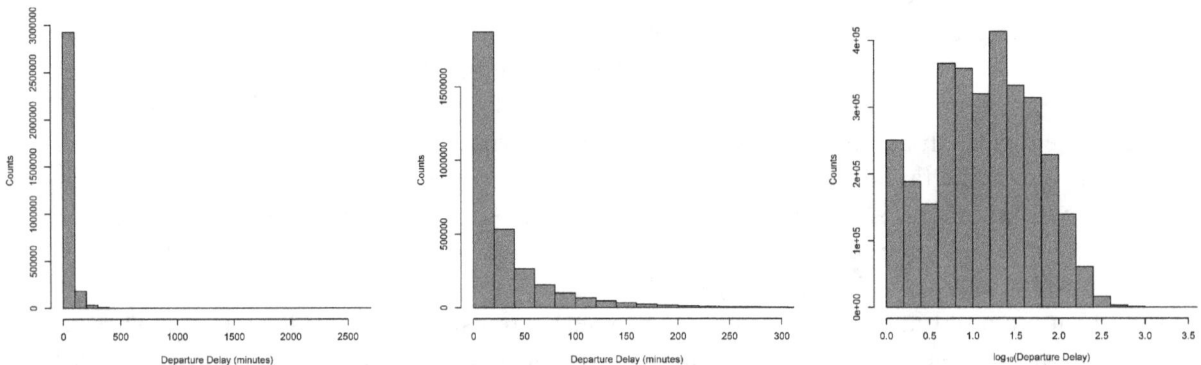

Figure 3.9: Histograms of 2007 departure delay times. The left histogram is for all of the data with delay times greater than zero; the middle histogram is for departure delays between 1 and 300 minutes; and, the right histogram is of the log-transformed departure delay times (for delay times greater than zero).

> **Exercise 3.2** Using the 2007 airline data, create a histogram showing the elapsed time of flight (which is contained in the *ActualElapsedTime* variable). ∎

3.2.3 Lattice (or Trellis) Plots

Lattice plots (also known as trellis plots) are an array of some type of statistical graph of one variable subset according to the values of one or more categorical variables. The categorical variables are also referred to as conditioning variables. The idea is to create a series of plots of one variable, say bar charts of the fraction of delayed flights in each cancellation category by separate levels of some categorical variable such as airline. Figure 3.10 is exactly this lattice plot for the 2007 airlines data. Here the conditioning variable is airline, of which there are 20, which results in a lattice of 20 bar charts.

What Figure 3.10 shows is that the distributions of cancellation reasons differs by airline. For example, while cancellation code A (carrier-related cancellation) is often the most frequent reason for flight cancellation, for some airlines (American, American Eagle, Continental, Comair, and Jet Blue) cancellation code B (weather-related cancellation) is the most frequent reason, and for only two airlines (Expressjet and Skywest) is cancellation code C (NAS-related cancellation) most frequent. Interestingly, for Aloha and Hawaiian airlines, all cancellations are carrier-related, presumably because on their routes weather and NAS are less of an issue than for other airlines.

∎ **Example 3.5 — R Code for the Lattice Plot in Figure 3.10.** The code below generates the histograms in Figure 3.10. It is substantially more complicated than most of the R code we have used thus far, so below the code is a line-by-line explanation.

```
> library(lattice)
> airline.names.alpha_order <- c("AirTrans","Alaska","Aloha","American",
    "American Eagle","Atlantic Southeast","Comair","Continental","Delta",
    "Expressjet","Frontier","Hawaiian","JetBlue","Mesa","Northwest",
    "Pinnacle","Skywest","Southwest","United","US Airways")
> airline.codes.matching.alpha_order <- c("FL","AS","AQ","AA","MQ","EV",
    "OH","CO","DL","XE","F9","HA","B6","YV","NW","9E","OO","WN","UA","US")
> UC <- factor(data07$UniqueCarrier,levels=airline.codes.matching.alpha_order,
    labels=airline.names.alpha_order)
> CancelCode <- data07$CancellationCode
> cnts <- table(CancelCode,UC)[2:5,]
> pp <- as.data.frame.table(prop.table(cnts,margin=2))
> barchart(pp$Freq ~ pp$CancelCode | pp$UC,xlab="Cancellation Code",
    ylab="Fraction")
```

The first line loads the `lattice` library so that we can ultimately use its `barchart` function in the last line. In the next two lines we manually create a vector of airline names in alphabetical order and a vector of the airline codes corresponding to the alphabetical names. We do this so that the final plot will use the airlines names rather than the codes, and so that the names will be in alphabetical order, all to make the plot more readable.

Now, in the fourth line, using the two vectors we just created, we create a new vector that directly corresponds to the *UniqueCarrier* data vector, but our new *UC* vector now has airline names rather than codes and it knows to print them out in alphabetical order. In the fifth line we simply extract the

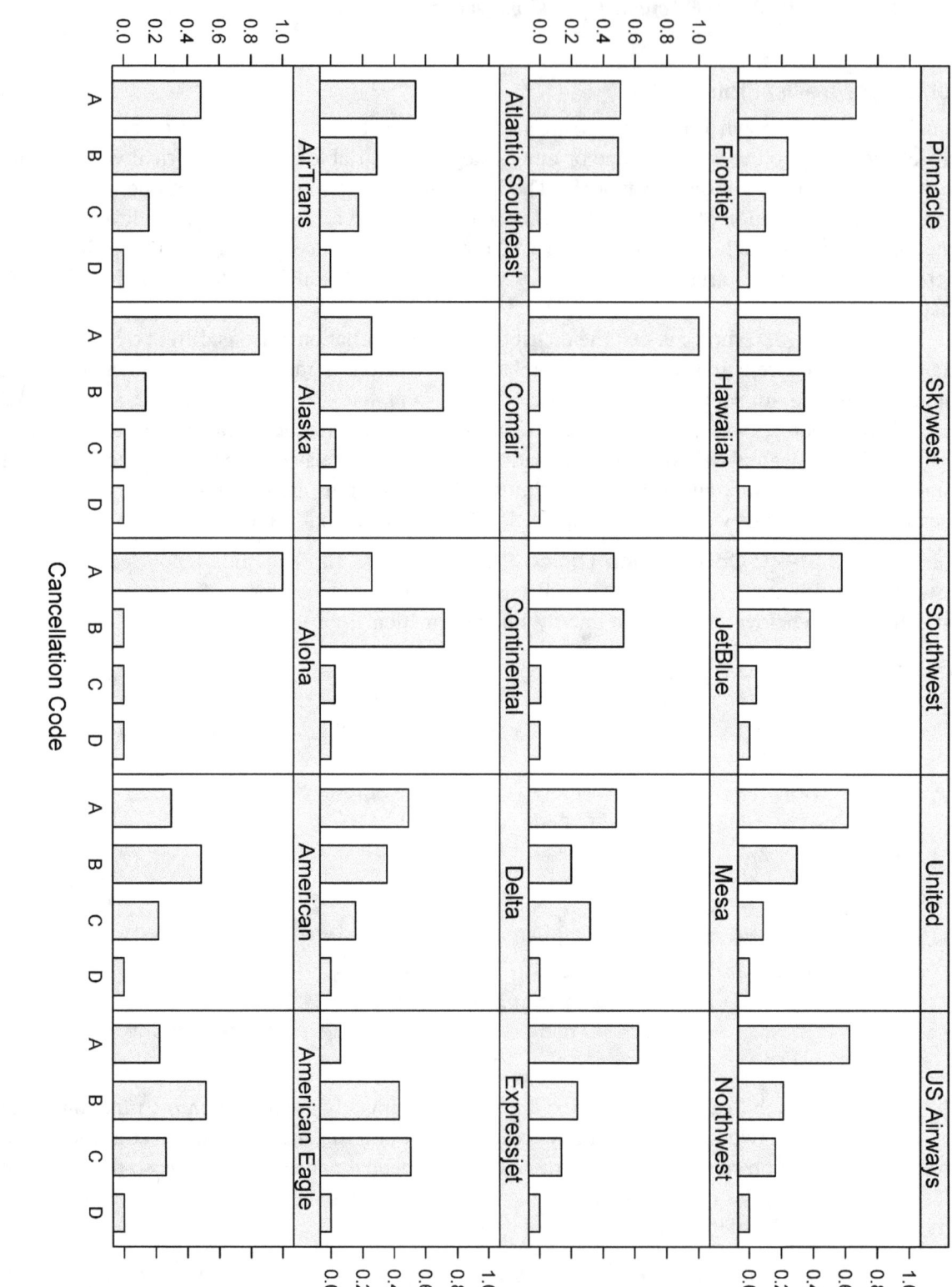

Figure 3.10: Lattice plot of the frequency of cancelled 2007 flights by cancellation code conditioned on airline.

data from the *CancellationCode* vector and put it in a new vector called *CancelCode*. In the sixth line we create a table of counts (called *cnts*) from the two vectors, but we exclude the first row of counts that are for those flights not cancelled. In the seventh line we create a new table called *pp* that converts the counts to fractions by airline.

Finally, with all that as set-up, in the last line we create the lattice plot. Besides the use of the barchart() function, the key part of the code is pp$Freq ~ pp$CancelCode | pp$UC. This says to plot the Freq variable for each category of the CancelCode variable and do this for each level of the UC variable. As you can see in Figure 3.10, this corresponds to plotting the fraction of cancelled flights for each of the cancellation categories by each of the airlines.

To really understand the code, sequentially run each line above and take a look at the resulting object. That will clearly show you what each step is doing and you'll understand both the output of each step and thus what the inputs are for the next step. ∎

In some ways, a lattice plot is nothing more than a set of repeated graphs, one for each category of the conditioning variable or set of categories of the conditioning variables. Three things make lattice plots more useful then simply manually repeating some plot by variable levels. The first is that good software facilitates exploring the data by making it easy to generate lattices. Second, and more importantly, the graphs in the lattice are all plotted with the horizontal and vertical axes on the same scale. This makes the plots easier to compare across the various categories. Third, the lattice can be conditioned on more than one categorical variable, which allows for the discovery of more complicated relationships in the data.

> **Exercise 3.3** Using the 2007 airline data, create a lattice plot of elapsed time of flight histograms by airline. Hint: Start by re-running the first four lines of code from Example 3.5. Then, use the histogram() function from the R lattice package, where you will be plotting the *ActualElapsedTime* variable conditioned on the *UC* variable using the syntax
>
> ```
> histogram(~data07$ActualElapsedTime | UC)
> ```
> ∎

3.2.4 Box Plots

Box plots are useful for depicting the distributions of continuous data. They do so by displaying summary statistics of the data, including the median and the lower and upper hinges (or first and third quartiles in some statistical software programs). As a result, box plots only require one dimension (unlike histograms that need two dimensions). Because the box plot is based on summary statistics, some information is lost, but box plots can still be very informative, particularly when comparing the distributions of two or more sets of data.

As shown in Figure 3.11, to construct a box plot first calculate the median and the hinges of a set of data. A box is then plotted that connects the upper and lower hinges (roughly, the 25th and 75th percentiles), and a line is added inside the box to show the median. At each end of the box *whiskers* are added by extending lines from the box that are 1-1/2 times the interquartile range (IQR). These lines are then truncated back to the last point contained within the line. Each whisker thus terminates at an actual data point, which means the whiskers will likely be of different lengths. Finally, observations that fall outside of the whiskers are indicated by dots and are designated outliers.

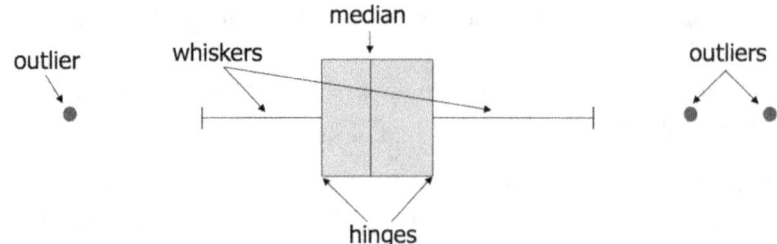

Figure 3.11: An illustration how a box plot is constructed from summary statistics. Box plots are useful for depicting the distribution of continuous data in one dimension.

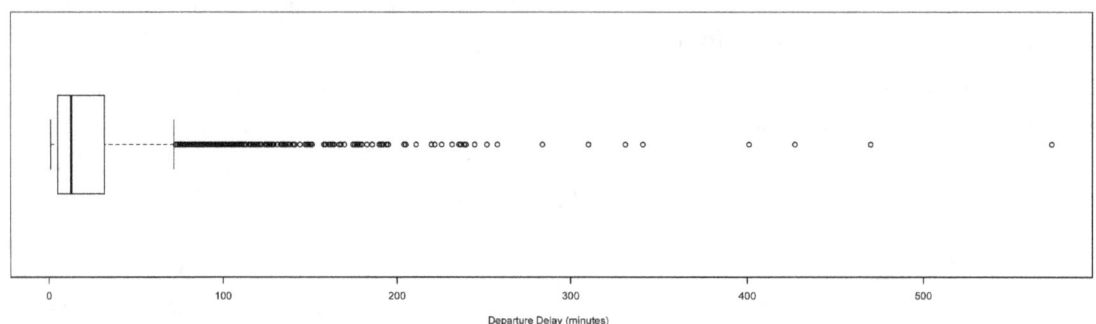

Figure 3.12: Example box plot: Departure delay for 2007 flights originating in Monterey, California, for flights that were delayed.

So, a box plot displays a lot of information: a measure of central tendency, the median; measures of how variable the data are as indicated by the width of the box (which is or is close to the IQR) and the length of the whiskers; and potentially unusual points, the outliers. Note that how box plots are drawn can vary by software package. For example, when drawing the whiskers some packages allow the user to choose multiples other than 1-1/2 times the IQR and some extend the whiskers all the way to the minimum and maximum values in the data.

For example, Figure 3.12 shows a (horizontal) box plot of the departure delay for 2007 flights originating in Monterey, California (for flights that were actually delayed). Focusing on the box, we see that most delayed flights were delayed for a relatively short period of time – roughly a half hour or less. However, the long right tail shows that some flights were delayed for very long times, with the longest delay being greater than 500 minutes (or more than 8 hours).

As with histograms, box plots can also get distorted and hard to read when the data are extremely skewed. Figure 3.12 is still reasonably readable, but a box plot equivalent to the the left histogram in Figure 3.9 would be equivalently distorted and thus difficult to read. However, in such cases, just like with histograms, transforming the data can make the plot easier to read. For example, Figure 3.13 is a box plot of the log-transformed 2007 departure delay data for Monterey flights. Compare Figure 3.13 to Figure 3.12.

The log base 10 transformation is particularly useful because it is easy to mentally convert back to the natural units: zero on the log base 10 scale is 1 on the natural scale, 1 is 10 on the natural scale, 2 is 100 on the natural scale, etc. Just think of integers in the log base 10 scale as counting the

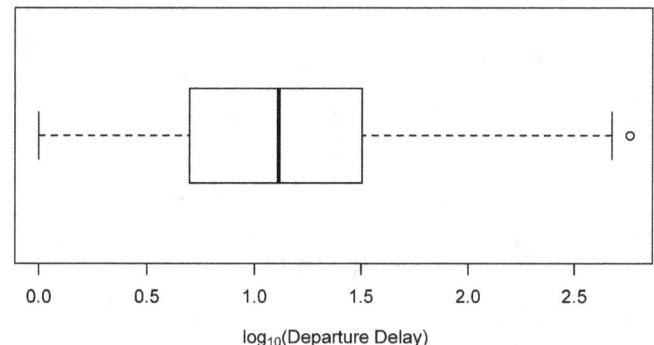

Figure 3.13: Box plot of the log (base 10) transformed departure delays for 2007 flights originating in Monterey, California, for flights that were delayed.

the zeros (with a one in front) on the natural scale.

■ **Example 3.6 — R Code for the Box Plots in Figures 3.12 and 3.13.** Using the *DepDelay* variable in the 2007 data, subset to only the Monterey flights using the *Origin* variable, the code below generates the box plots in Figures 3.12 and 3.13.

```
> boxplot(data07$DepDelay[data07$Origin=="MRY" & data07$DepDelay > 0],
  xlab="Departure Delay (minutes)",horizontal=TRUE)
> boxplot(log10(data07$DepDelay[data07$Origin=="MRY" & data07$DepDelay > 0]),
  xlab=expression(paste(log[10],"(Departure Delay)",sep="")),horizontal=TRUE)
```

The `horizontal=TRUE` option plots the box plots horizontally; the default is to plot them vertically. The use of the `expression()` function with the `xlab` option allows for writing mathematical expressions on the axes labels.   ■

A very useful way to compare observations grouped by a categorical variable is via side-by-side box plots where, similar to lattice plots, a separate box plot is created for each category. Thus, side-by-side box plots require a continuous and a categorical measure on each observation; for example, departure delay and airline. Also like lattice plots, one of the reasons side-by-side box plots are powerful is that all the box plots are graphed on the same scale which facilitates comparisons between the categories.

For example, Figure 3.14 is a side-by-side box plot of the $\log_{10}$-transformed departure delays for 2007 by airline. The figure clearly shows which airlines had better departure delay performance (meaning their departure delay distributions that tend towards smaller delays) – e.g., Hawaiian – and which airlines had worse departure delay performance (meaning their departure delay distributions tended towards larger delays) – e.g., Mesa.

■ **Example 3.7 — Creating side-by-side box plots in R.** Using the *UC* vector created for the lattice plot in Example 3.5, the code below generates the plot in Figure 3.14.

```
> par(mar = c(9,5,4,4))
```

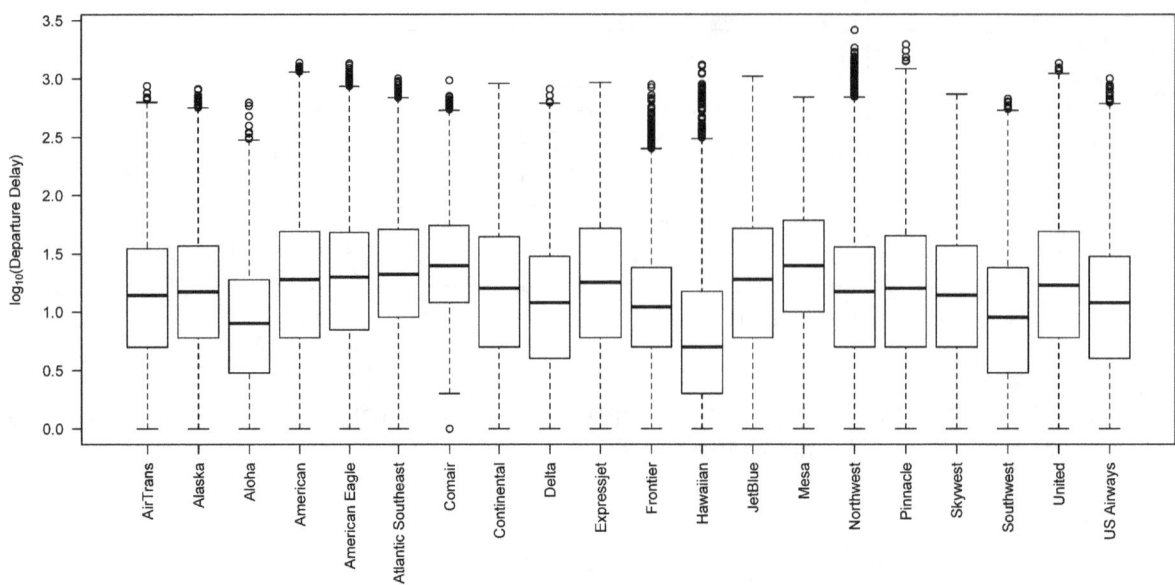

Figure 3.14: Example side-by-side box plot: Plot of the $\log_{10}$-transformed departure delays for 2007 by airline.

```
> boxplot(log10(data07$DepDelay[data07$DepDelay>0])~UC[data07$DepDelay>0],
  ylab=expression(paste(log[10],"(Departure Delay)",sep="")),las=2)
```

First note that we're using the same `boxplot()` function as we did in Example 3.6. The difference here is in the arguments, particularly the syntax

$$\log10(data07\$DepDelay[data07\$DepDelay>0])\sim UC[data07\$DepDelay>0].$$

What this tells the `boxplot()` function to do is to plot the $\log_{10}$-transformed departure delay values (for those observations with positive departure delays) as a function of the *UC* vector (again, only for those UC values corresponding to flights with positive departure delays).[2]   That is, this syntax tells R to create a separate box plot for each unique value in the *UC* vector, where note that this only makes sense if the *UC* vector is categorical.

Second, the `las=2` argument tells the `boxplot()` function to rotate the labels (in this case, the airline names) 90 degrees so that they are readable. (Run the code without this argument and see what you get.) However, because some of the names are long, we also need to increase the lower margin of the plot so that the names all fit. That is done in the first line, where the `par()` function controls lots of graphical parameters. The option `mar` is for adjusting the margins, where the vector sets the margins starting at the bottom and working clockwise around the plot. (Note that the margins remain at the specified setting until either the `par()` function is re-run with the `mar` option or the plotting window is closed.)   ■

---

[2]The $\sim$ symbol is read and means "as a function of." We will use this notation more in later chapters.

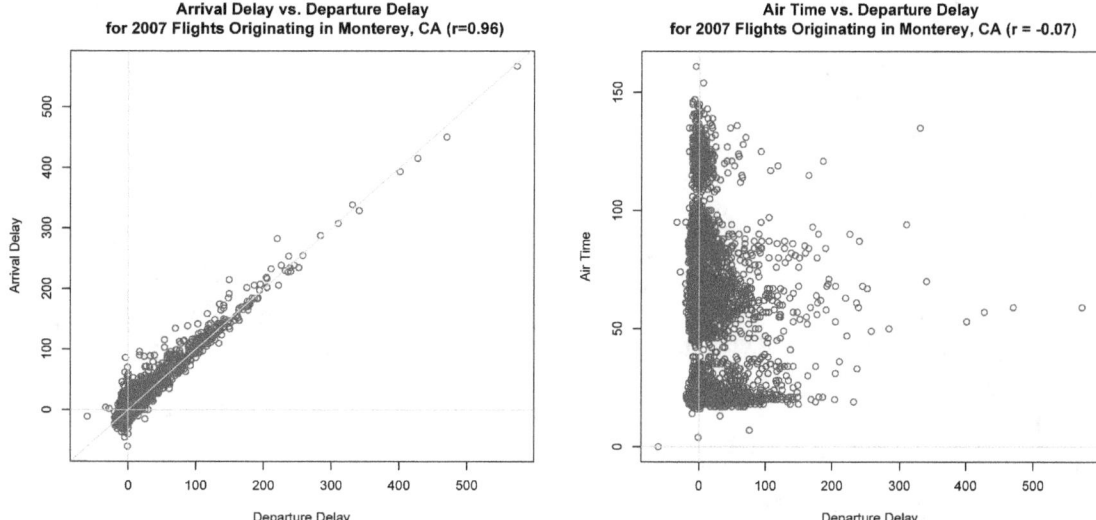

Figure 3.15: Example scatterplots: For all 2007 flights originating from the Monterey, California airport, arrival delay (in minutes) versus departure delay (in minutes) on the left and flight air time (in minutes) versus departure delay on the right.

> **Exercise 3.4** Using the 2007 airline data, create a box plot of elapsed time of flight for all the flights. Then create side-by-side box plots of elapsed time of flight by airline (using the *UC* variable). ∎

3.2.5 Scatterplots

A scatterplot is a graph of one continuous variable versus another. Scatterplots are useful for summarizing data and they are particularly effective at showing whether there is a relationship between two variables, where such a relationship would show up as a pattern in the plot. Anscombe's data in Section 3.1 is the canonical example of how scatterplots can reveal patterns that might otherwise go unnoticed.

Figure 3.15 shows two examples of scatterplots, where the left plot is a scatterplot of arrival delay versus departure delay for all the 2007 flights originating from the Monterey, California airport. What the scatterplot shows, perhaps not surprisingly, is that there is a high correlation ($r = 0.96$) between the two variables where, presumably, departure delays cause arrival delays. The plot also shows that there are some flights with little or no departure delay that do have arrival delays, probably because of weather and air traffic congestion encountered enroute. However, the plot makes it quite clear that the largest driver of arrival delay is departure delay.

The right scatterplot in Figure 3.15 is another example, plotting flight air time versus departure delay for all of the 2007 flights originating from the Monterey airport. Here we see that there are clusters of what look like four groupings of flights that take roughly the same air time. The scatterplot shows that these air times are essentially (and perhaps unsurprisingly) uncorrelated with departure delay ($r = -0.07$). That is, the plot shows that, once a flight is airborne, its flight time is not affected by the length of departure delay.

∎ **Example 3.8 — R code for the scatterplots in Figure 3.15.** Using the *DepDelay*, *ArrDelay*, and *AirTime* vectors from the 2007 data, the code below generates the scatterplots in Figure 3.15.

```
> plot(data07$DepDelay[data07$Origin=="MRY"],data07$ArrDelay[data07$Origin==
  "MRY"],xlab="Departure Delay",ylab="Arrival Delay",main="Arrival Delay vs.
  Departure Delay\n for 2007 Flights Originating in Monterey, CA (r=0.96)",
  col="blue")
> lines(c(-500,3000),c(0,0),col="gray")
> lines(c(0,0),c(-500,3000),col="gray")
> lines(c(-500,4000),c(-500,4000),col="gray")
> plot(data07$DepDelay[data07$Origin=="MRY"],data07$AirTime[data07$Origin==
  "MRY"],xlab="Departure Delay",ylab="Air Time",main="Air Time vs.
  Departure Delay\n for 2007 Flights Originating in Monterey, CA (r=-0.07)",
  col="blue")
> lines(c(-500,3000),c(0,0),col="gray")
> lines(c(0,0),c(-500,3000),col="gray")
```

While the code looks complicated, the basic syntax for a scatterplot of vectors x and y is `plot(`x`,` y`)`. The `lines()` function overlays a line on the plot, where the line connects the two points $\{x_1, y_1\}$ and $\{x_2, y_2\}$ via the syntax `lines(c(`x_1, x_2`),c(`y_1, y_2`))`. ∎

Now, we might want to check on the hypothesis that, in fact, there are four groups of flight times. Table 3.1 shows that flights from Monterey went to 11 different destinations in 2007 (where the vast majority went to Los Angeles and San Francisco).

Destination	DEN	FAT	LAS	LAX	LGB	ONT	PHX	PIH	SAN	SFO	SLC
# Flights	364	1	140	3,513	101	413	714	1	464	2,475	582

Table 3.1: Counts of all 2007 flights originating from Monterey, California, by destination airport.

Digging a bit deeper, the side-by-side box plots in Figure 3.16 show that there are either five or six groups of similar flight times. For example, Figure 3.16 shows that the flight times to Las Vegas ("LAS") and Los Angeles ("LAX") are similar, as are the flight times to Long Beach ("LGB") and Ontario ("ONT") airports, and as are the flight times to Phoenix ("PHX") and Salt Lake City ("SLC"). Flight times to Denver ("DEN") and San Francisco ("SFO") are clearly different from the other destinations, though if we look a bit more closely we see that sometimes it takes as long to fly to Phoenix as to go to Denver, and San Diego ("SAN") flight times are somewhere in between LAS/LAX and LGB/ONT times.

> **Exercise 3.5** Using the 2007 airline data, create a scatterplot of elapsed time of flight versus flight air time (*AirTime*) for all the flights. Now only plot those flights that originating from Dallas Love Field airport (`data07$Origin=="DAL"`). ∎

When there are more than two continuous variables, a *scatterplot matrix* is useful for displaying the data. A scatterplot matrix uses a lattice-like format to efficiently show scatterplots for all pairs of variables. For example, Figure 3.17 shows a scatterplot matrix of *AirTime*, *Distance*, *ArrDelay*, and *DepDelay* data for the 2007 Monterey flights.

To read the plot, first note that the diagonal identifies the variables associated with the axes of each of the individual scatterplots. Thus, for example, *ArrDelay* is the variable on the y-axis of all

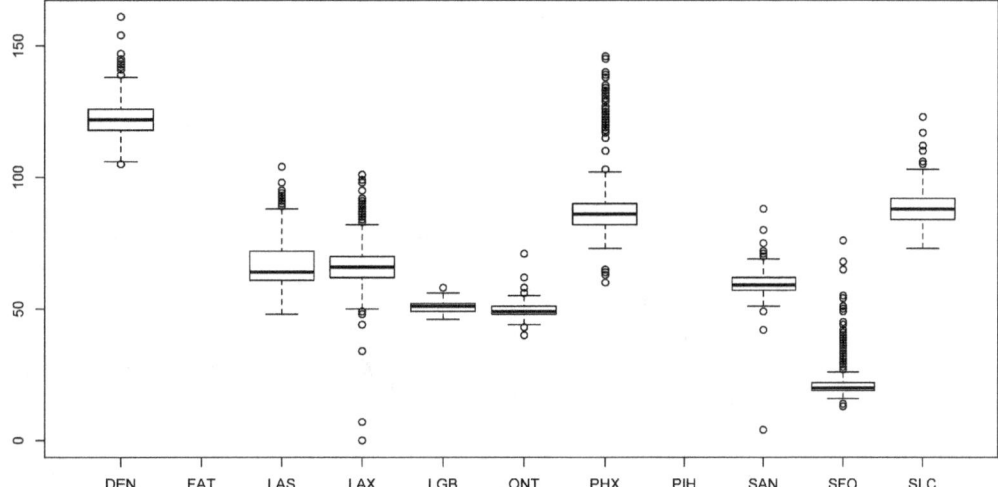

Figure 3.16: Side-by-side box plots of flight air time (in minutes) by destination airport for all the 2007 flights originating from the Monterey, California airport.

the plots in the top row and it is the variable on the *x*-axis of all the plots down the left column. So, for example, the left-most scatterplot in top row is a scatterplot of *ArrDelay* versus *DepDelay* – note that it matches the left scatterplot in Figure 3.15. Similarly, the plot in the third row of the second column is a scatterplot of *AirTime* versus *DepDelay* and it matches the right scatterplot in Figure 3.15.

From this particular scatterplot matrix, we can learn a lot about the 2007 Monterey flights. For example, as we first learned in Figure 3.15, here we see that departure and arrival delays are highly correlated while flight time looks independent of both departure and arrival delays. We also see that the distances flown out of the airport are very discrete with only five or six unique values. Yet, we also see that for a given distance the flight time can vary quite a bit.

Also note how each pair of variables is plotted twice but with the axes reversed. In fact, all the plots above the diagonal have a matching reversed plot below the diagonal, so it is really only necessary to examine half of the plots in the matrix. In fact, a scatterplot matrix of *k* variables has $k(k-1)/2$ unique plots.

Example 3.9 shows the code for generating the scatterplot matrix in Figure 3.17. Because the `pairs()` function was run on four variables, it produced a 4×4 matrix, but you can create any size matrix. The only limiting factor is the size of the individual scatterplots where, with too many variables, the plots get too small to be readable.

■ **Example 3.9 — R code for the scatterplot matrix in Figure 3.17.** The `pairs()` function run on a matrix or dataframe with more than two continuous variables produces a scatterplot matrix. In the code below, it was run on the 14th, 15th, 16th, and 19th columns of `data07` dataframe (which correspond to the *AirTime*, *ArrDelay*, *DepDelay*, and *Distance* variables), and only for those rows in which the *Origin* variable is equal to "MRY."

```
> pairs(data07[data07$Origin=="MRY",c(15,16,14,19)],col="blue")
```
■

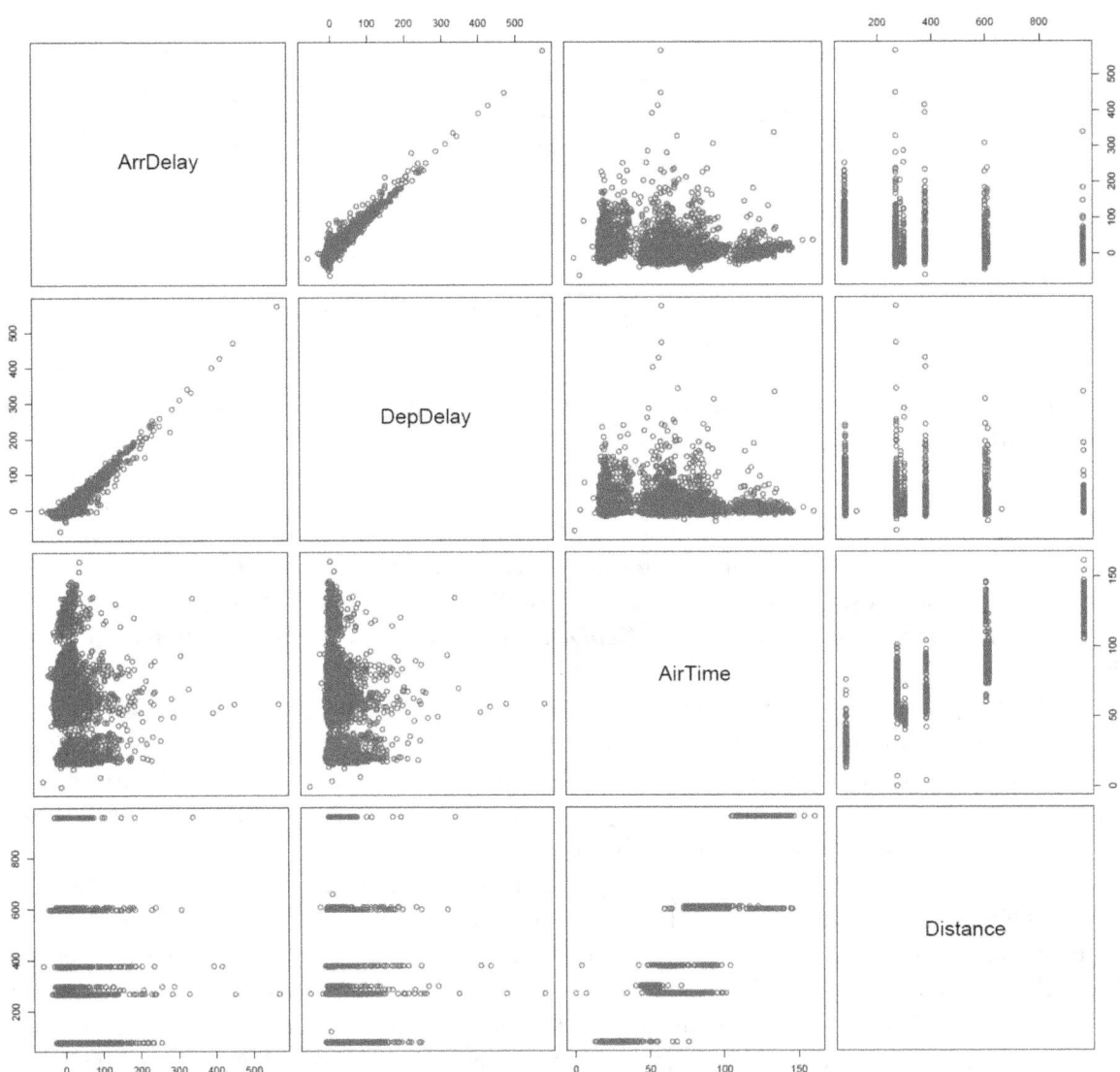

Figure 3.17: Scatterplot matrix of *ArrDelay*, *DepDelay*, *AirTime*, and *Distance* variables for all 2007 flights originating from Monterey, California.

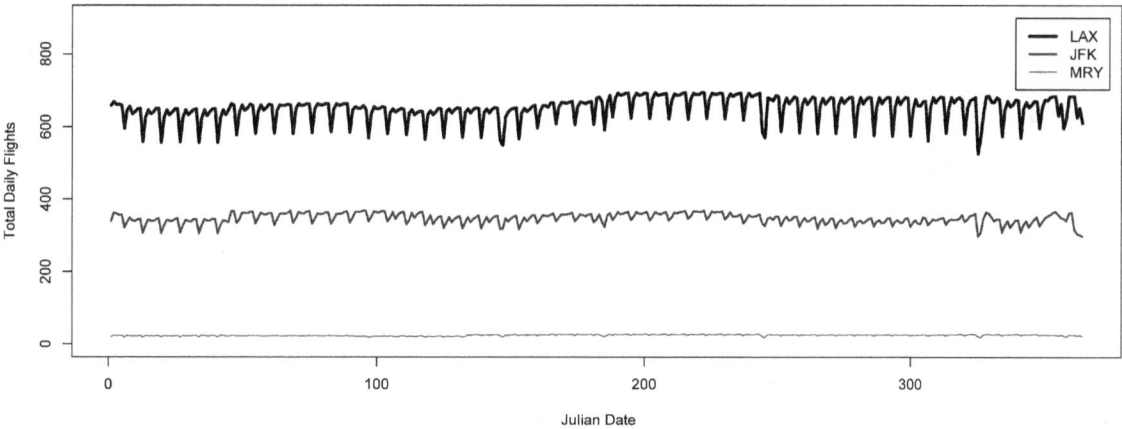

Figure 3.18: Time series plot of the total number of daily flights out of LAX (Los Angeles airport), JFK (John F. Kennedy airport), and MRY (Monterey airport) in 2007. The *Julian Date* is just the days of the year sequentially numbers, where 1 corresponds to January 1st and December 31st is either 365 or 366 depending on whether it is a leap year or not.

3.3 Graphics for Longitudinal Data

Graphics for longitudinal data are in many ways similar to those for cross-sectional data. (See Chapter 2 (page 26) for additional discussion about types of data.) A key difference, however, is that time is an important variable that must be displayed in some useful and informative way. For many longitudinal graphics, that often means that the variable plotted on x-axis of the plot is time.

3.3.1 Time Series Plots

Time series are data collected or measured repeatedly over time, where the term "time series" is often used as a synonym for longitudinal data. However, time series *plots* display longitudinal data with respect to time. By convention, the data are plotted with the magnitude of the observation on the y-axis and time, in the appropriate units, on the x-axis. Time series plots are useful for assessing whether there are trends in the data, such as a regular increase in the number of flights departing from an airport (perhaps resulting from population growth in the region served by the airport), or whether there are cycles in the data (perhaps as a result of weekly, seasonal or other influences).

For example, Figure 3.18 is a time series plot of the total number of daily flights out of LAX (Los Angeles airport), JFK (John F. Kennedy airport), and MRY (Monterey airport). The x-axis contains the sequential days of the year (referred to as the *Julian Date*, where 1 corresponds to January 1st and December 31st is either 365 or 366 depending on whether it is not or is a leap year). The y-axis is the total number of flights on each day of the year.

Note that it is convention in time series and other longitudinal plots to display the data using lines that connect what otherwise would be discrete points on the plot. That is, with this data for each day and each airport there is a daily count. While we could have plotted these data as points in a scatterplot-like display, much like the plots in Section 3.2.5, by using lines to connect the data it helps make the time trends visually clear and it also helps to communicate to the reader that this is a time series plot and not a scatterplot. Note, for example, how the lines in Figure 3.18 show a

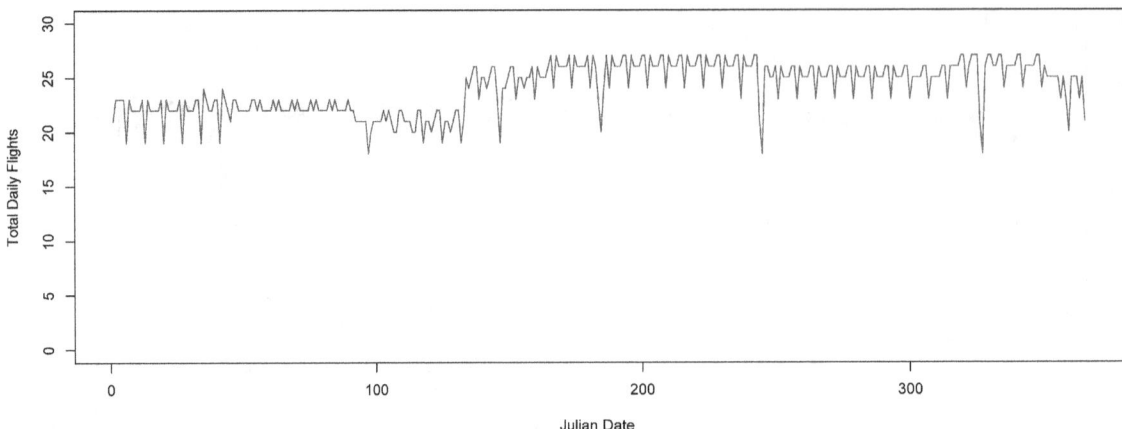

Figure 3.19: Time series plot of the total number of daily flights just out of MRY in 2007, where the *y*-axis scale now makes the weekly cycle much more apparent.

clear weekly cycle in the number of flights, where the regular dips in the number of flights occur on Fridays. We also see a very slight increase in the number of flights in the summer, at least in LAX, and less regularity in the weekly pattern towards the end of the year during the holidays.

Figure 3.19 shows only the Monterey data, where the *y*-axis scale now makes the weekly cycle much more apparent for this data. Also, the discrete nature of the data is more obvious in this plot, as well as what may be some seasonal effects. For example, we see a slight increase in the number of summer flights, and the major holidays are visible as downward spikes: the day before Memorial Day, the day before Labor Day, Thanksgiving and the day after, Christmas, and New Years Eve (corresponding to Julian dates 148, 244, 325 & 326, 359, and 365).

3.3.2 Repeated Cross-sectional Plots

Another way to present time series data is via a series of repeated plots, where each plot represents a different (but sequential) time period. This is simply the graphical equivalent to the approach in Section 2.3 for summary statistics. For example, box plots repeated over time can be be very informative. Figure 3.20 shows this approach for the LAX departure delay data, where each box plot is based on a month of data.

Remember that the horizontal line in the middle of a box plot is the median, and in Figure 3.20 the horizontal line across the eintire plot corresponds to the median departure delay for the entire year. Comparing these, we see that there was a cycle in median departure delays with greater median delays in the winter months (December, January, and February) and lower median delays in the spring and fall. Now, note that the *y*-axis on this plot is on a log base 10 scale, so actual differences are larger than the visually appear in the plot. For example, the median departure delay for all 2007 flights was 11 minutes while in December it was 16 minutes (i.e., $10^{1.20412} = 16$).

Lattice plots can also be used to display repeated cross-sectional plots if the conditioning variable corresponds to time. For example, Figure 3.21 is a lattice plot of of the cancellation codes for all of the flights cancelled in 2007. While this type of plot is not as useful as those in Figures 3.18 through 3.20 for showing trends, it does show that there are very different cancellation patterns and numbers

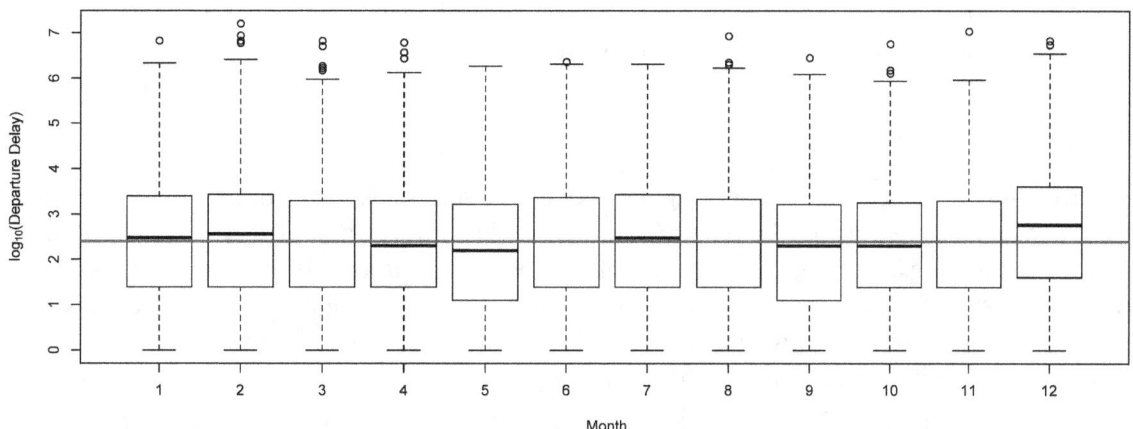

Figure 3.20: Repeated boxplots by month of departure delay (for those flights that were delayed) for 2007 LAX data.

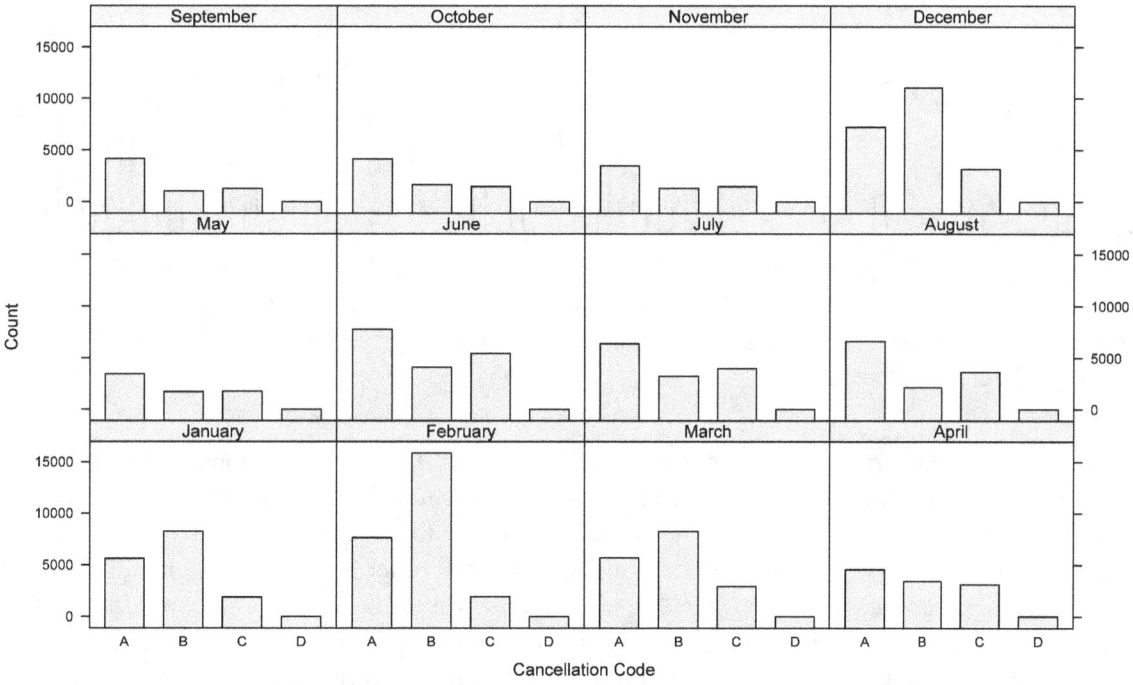

Figure 3.21: Lattice plot of cancellation codes by month for all of the flights cancelled in 2007.

of flight cancellations throughout the year. For example, we see that in December, January, February, and March, weather-related cancellations are the most frequent cause (particularly in February of 2007). However, throughout the rest of the year, carrier-related cancellations are the most frequent cause. Also, the plot clearly shows that there were more cancellations in the winter months (followed by the summer months) than in other times of the year

3.3.3 Autocorrelation Plots

Autocorrelation plots are used for checking whether a time series has trends or cycles. As the name suggests, an *autocorrelation plot* (also referred to as *correlogram*) is a plot of sample autocorrelation values for various lags, starting at $k = 1$ through whatever lag value is useful. (See Section 2.3.2, starting on page 49, for the definition of autocorrelation.) If the data do not have cycles or tends then all of the autocorrelations should be near zero for all of the lags. If there are cycles or trends then one or more of the autocorrelations will have a large positive or large negative value.

For example, Figure 3.22 shows the autocorrelation plots for the number of daily flights from LAX, JFK, and MRY for 2007 (first shown in the time series plots in Figures 3.18 and 3.19). "Large" autocorrelations are those with bars that fall outside the two dotted lines (where the "ACF" label on the y-axis label stands for *autocorrelation function*). Those inside the dotted lines can be considered close enough to zero that they can be ignored.[3]

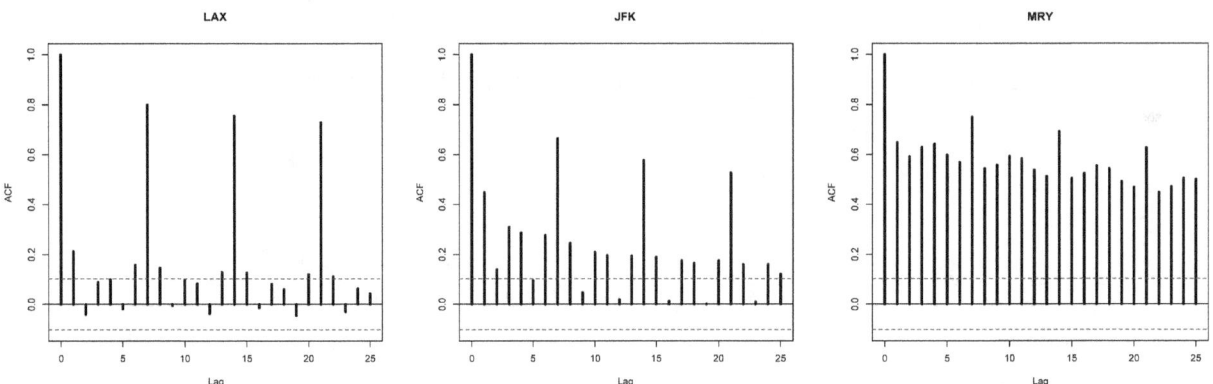

Figure 3.22: Autocorrelation plots for the number of daily flights from LAX, JFK, and MRY for 2007 (see Figures 3.18 and 3.19).

So, the large autocorrelation values in the LAX plot at lags 0, 7, 14, and 21 indicate that there is a strong weekly cycle in the data (which we also observed in Figure 3.18). The JFK plot has similarly large autocorrelation values at lags 0, 7, 14, and 21, also indicative of a strong weekly cycle. In addition, that plot shows decreasing autocorrelation values at other lag values which indicates that the JFK data also have a weak trend. Finally, the MRY data show both large autocorrelation values at lags 0, 7, 14, and 21 as well as generally large but decreasing values at all of the other lags which indicates both a weekly cycle as well as other trends in the data. These were evident in Figure 3.19 where we saw a number of different patterns throughout the year.

---

[3]That is, the autocorrelations within the dotted lines are "not statistically different from zero." We'll learn more about this when we get to later chapters on hypothesis testing.

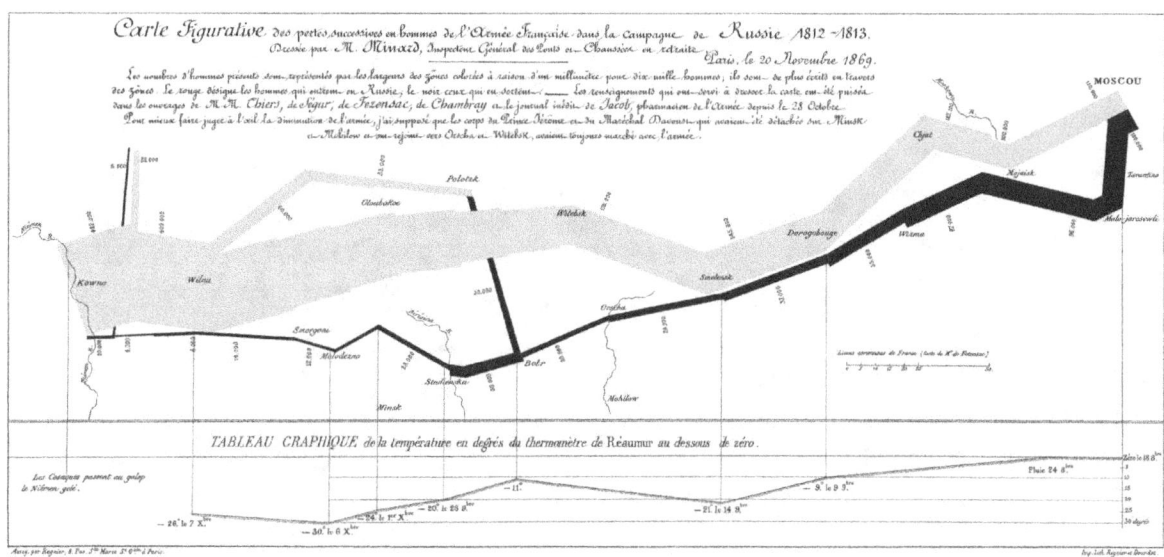

Figure 3.23: Classic graphic from Charles Minard (1781-1870) showing the progress and fate of Napoleon's army in its march against Moscow. This is a combination of a data map and a time-series, and displays six variables.

3.4 Other Types of Data Visualization

As we discussed in Section 3.2, traditional statistical graphics were mainly designed for displaying numerical data. Today, there are many different types of data that are readily available and that data scientists are interested in analyzing. These include:

- Social media and other types of text-based data;
- Surveys and related types of data;
- Geo-spatial and geographic information system (GIS) data;
- Infrastructure, social, and other types of network data; and,
- Video- and audio-based data.

These data may require a variety of cross-sectional, temporal, and spatio-temporal visualization analytical methods and visualization techniques. Furthermore, good visualization is optimized for the particular application and user. For example, methods for best visualizing networks may be quite different than those for displaying GIS data. Similarly, GIS methods for displaying point data on maps are not appropriate for displaying areal data from surveys. On the other hand, some types of social media data can be effectively displayed as a network while other types of social data will require other visualization methods.

One of the most famous and arguably one of the most elegant portrayals of multi-dimensional data is Charles Minard's flow map of Napoleon's March to and from Moscow. As shown in Figure 3.23, this graph simultaneously displays six different variables: army size, location (roughly) in two dimensions, direction of movement, and temperature during the army's retreat from Moscow all versus time.

Now, the number of possible visualization techniques, methods, and examples far exceeds what we can (or would want to) cover in this chapter. Instead, what we'll do is discuss a few visualization techniques just to illustrate the variety of possibilities

Figure 3.24: A word cloud visualizations of text data. The size and hue of a word is proportional to the number of times it appears. These word clouds were created using the R `wordcloud` package.

3.4.1 Text Visualization

One visualization specific to linguistic data is the word cloud (or "wordle") which is used as a visual summary of the words in a set of text. Words that appear more frequently in the text appear more prominently in the cloud; the cloud-building software arranges placement and orientation.

For example, the left word cloud in Figure 3.24 was created from the the text of *Green Eggs and Ham* by Dr. Seuss. Note how much of the story comes through, particularly the use of the word 'like,' but also note how because the word cloud is focused on the individual words, so the phrase 'do not like' that is used so often in the text is not visible. Similarly, there is an emphasis on the word 'will' that misses the frequent use of 'will not' in the text.

Another example is the right word cloud in Figure 3.24 that was created from the 2007 flight data using the airport origination airport codes. In it, the size of the airport code is proportional to the number of flights originating from that airport in 2007, where the word cloud clearly shows the major U.S. airports surrounded by a multitude of smaller airports.

In R, the `wordcloud` package can be used to create word clouds. Using other methods, word clouds can be made three dimensional, rotating, and/or interactive when they are rendered in some sort of electronic medium. Of course, no formal analyses are associated with a word cloud: the cloud itself is the end result.

3.4.2 Survey Data Visualization

The visualization of survey data has changed little in the past few decades. One challenge is that survey data are usually discrete and often ordinal, typically resulting from responses to *Likert scale* questions. A Likert scale is one in which a survey respondent is asked how much he or she agrees with a particular statement and then is given a response scale of the form "strongly agree," "agree," "neither agree nor disagree," "disagree," and "strongly disagree."

Standard displays of such data include pie and bar charts, though these types of graphics do not

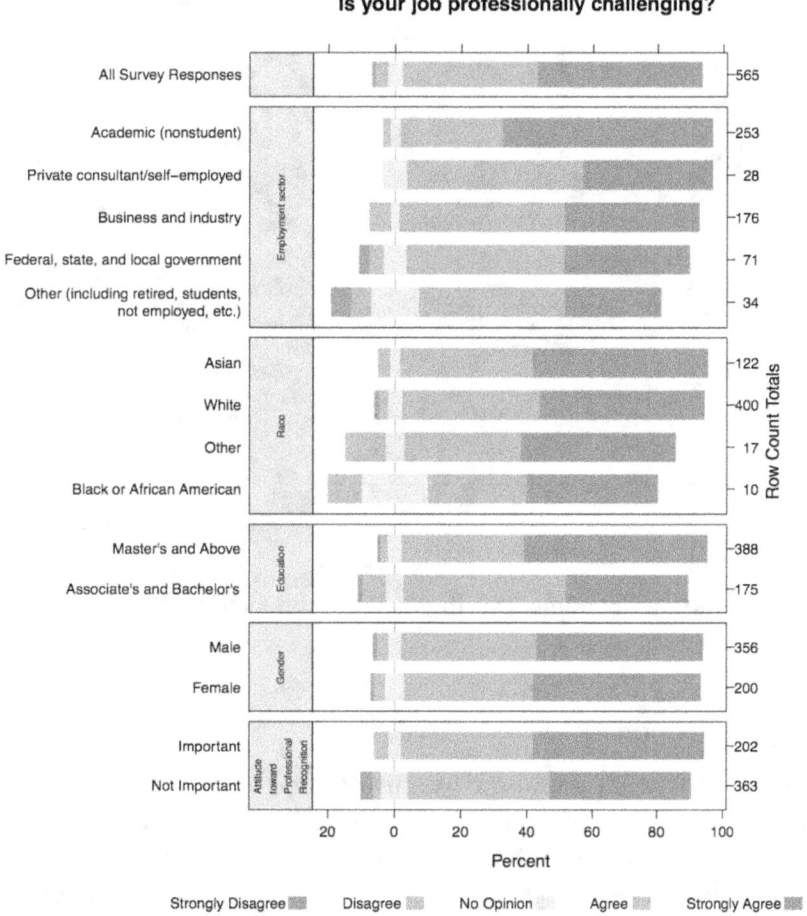

Figure 3.25: An example of a diverging stacked bar chart for a five point Likert scale question.[5] The top bar shows the entire sample. Underneath it are bars for subsets of the data associated with five different demographic categories.

lend themselves readily to displaying ordinal data. Also, these types of graphs are not particularly useful for displaying interactions between variables or relationships among multiple variables. Even using side-by-side and stacked bar charts, it is generally difficult to compare between subsets of the data.

A recent advance in survey data visualization is the diverging stacked bar chart. As shown in Figure 3.25, the diverging stacked bar chart centers the stacked bars, typically on the neutral or central response and then each end of the bar diverges away from the neutral. These types of plots make comparing response distributions between subsets of the data relatively easy, clear, and intuitive.

[5]Source: Heiberger, R.M., and N.B. Robbins (2014). Design of Diverging Stacked Bar Charts for Likert Scales and Other Applications, *Journal of Statistical Software*, **54**:5, 1–32. Available on-line at www.jstatsoft.org/v57/i05/paper.

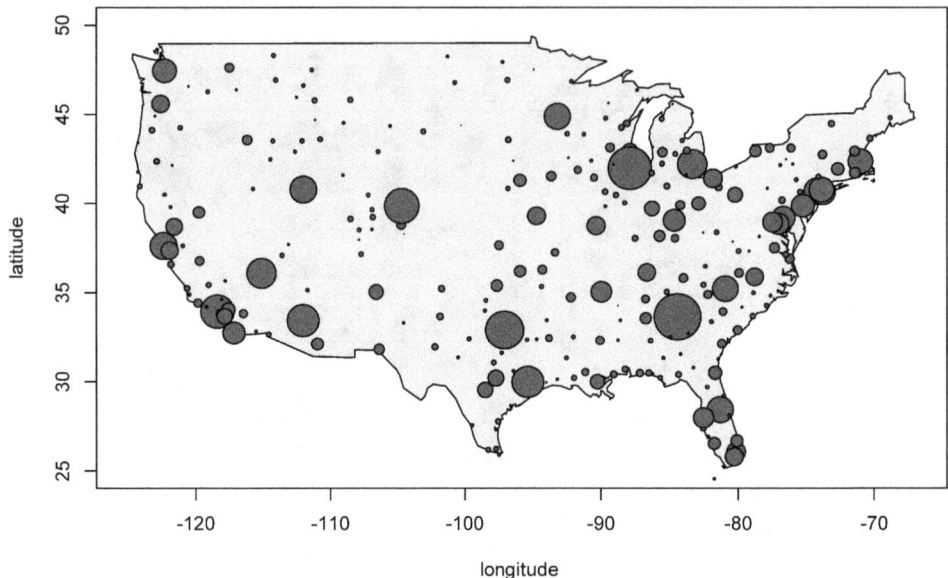

Figure 3.26: Bubble plot of the number of flights that originated at each domestic airport in the 48 continuous states in 2007.

3.4.3 Geo-spatial Visualization

The visualization of geographic data – the making of maps – has a long history. Developments in computing, though, have put the display of geographic data within easy reach: everyone can now make and plot maps using Google, Yahoo, MapQuest, Bing and similar software. Furthermore, the widespread incorporation of global positioning systems in devices such as mobile phones and freight palettes has led to significant growth in the amount of geo-referenced data available to consumers and organizations.

Typically, geo-referenced data are plotted on maps as points. This works well when the goal is simply to show the location of a site or sites on a map, but if the goal is to communicate numerical information associated with the points on the map then other graphical approaches are necessary. One solution is to use a shape centered at each location, the size of which is proportional to the numerical measure for the location. For example, Figure 3.26 is a *bubble plot* of the number of flights that originated at each airport in the flight data in 2007. Each circle ("bubble") is centered at each airport's location and the area of each circle is proportional to the number of flights 2007 flights originating at each airport.

Now, while the airport data can be associated with a specific point on a map, some data can only be associated with an area or region. Mapping such *areal data* can visually distort the information being conveyed if the associated areas differ in size. For example, the left plot in Figure **??** is a map showing the states, plus the District of Columbia, whose electoral votes were awarded to John McCain and to Barack Obama in the 2008 U.S. presidential election. About 62% of the map is shaded for McCain, but equal areas do not denote equal densities of electoral votes: Alaska's 570,000 square miles of land are much more visible than D.C.'s 68, though each contributes the same number of electoral votes (three).

One interesting technique to address this is the *cartogram*, in which a "map" shows regions that are scaled to represent the quantity being displayed, while having their respective shapes retained

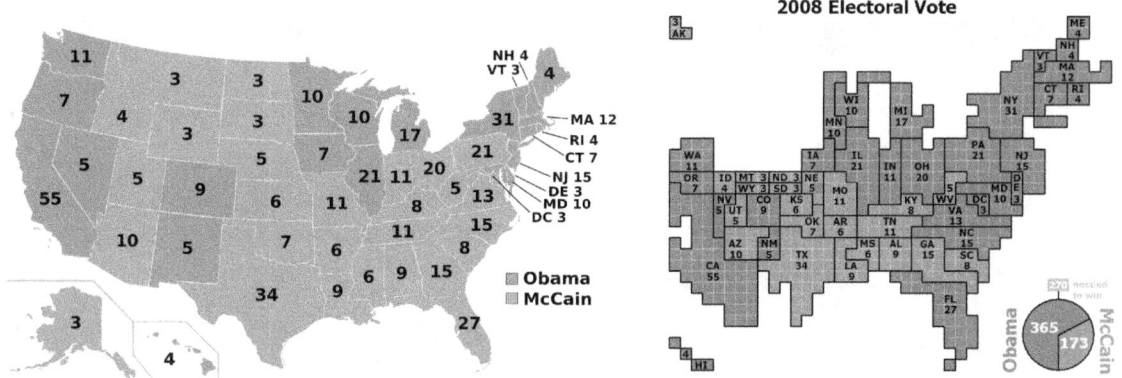

Figure 3.27: Left: Traditional map display of the states whose electoral votes were awarded to John McCain and to Barack Obama, plus the District of Columbia. Only about 38% of the map is shaded in favor of Obama, but he won 67.8% of the electoral college votes (365 out of 538). Right: In this *cartogram*, states are represented by areas proportional to the number of electoral college votes to which they are entitled and about 68% of the map is shaded for Obama.[7]

to the extent possible. In the cartogram to the right in Figure 3.27, the data and color-coding are the same the map on the left, but each state is represented by an area proportional to its number of electoral votes. Visually, this is a much truer representation of the data, where this map is about 68% shaded for Obama and Obama won 67.8% of the electoral college votes (365 out of 538). Cartograms can also be drawn with curved boundaries.

3.4.4 Network Visualization

Networks occur naturally in many situations in which there are entities (often called nodes), pairs of which are connected in some manner (where the linkages between entities are often called arcs). One example is a computer network, where each computer is a node, and where the connections between pairs of computers are arcs. Another example is electronic communication where people are the nodes and the arcs are between those pairs of people who communicate, perhaps via e-mail, or twitter, or on some social media platform.

While the idea of a network may be conceptually simple, visualizing networks in ways useful for learning about and understanding the network may be anything but simple. For example, particularly with large networks, simply displaying all the arcs and nodes results in what is appropriately referred to as a "hairball." For example, Figure 3.28 is a network "visualization" of all of the 2007 flights between United States airports. With 310 nodes (airports) and 5,736 arcs, this graphic is so complicated and busy that almost nothing about the network can be discerned except there are lots of nodes and arcs.

At issue is that hairballs fail to provide insight into the network other than to show that the displayed network is complex. Also, the idea that showing an entire network is insightful is akin to thinking that showing someone an entire dataset is a useful descriptive statistics strategy. Instead, it

---

[7]Sources: "Althistory Wiki" at http://althistory.wikia.com/wiki/United_States_presidential_election_of_2008_(SIADD) and "Cartogram-2008 Electoral Vote" from http://en.wikipedia.org/wiki/File:Cartogram-2008_Electoral_Vote.png.

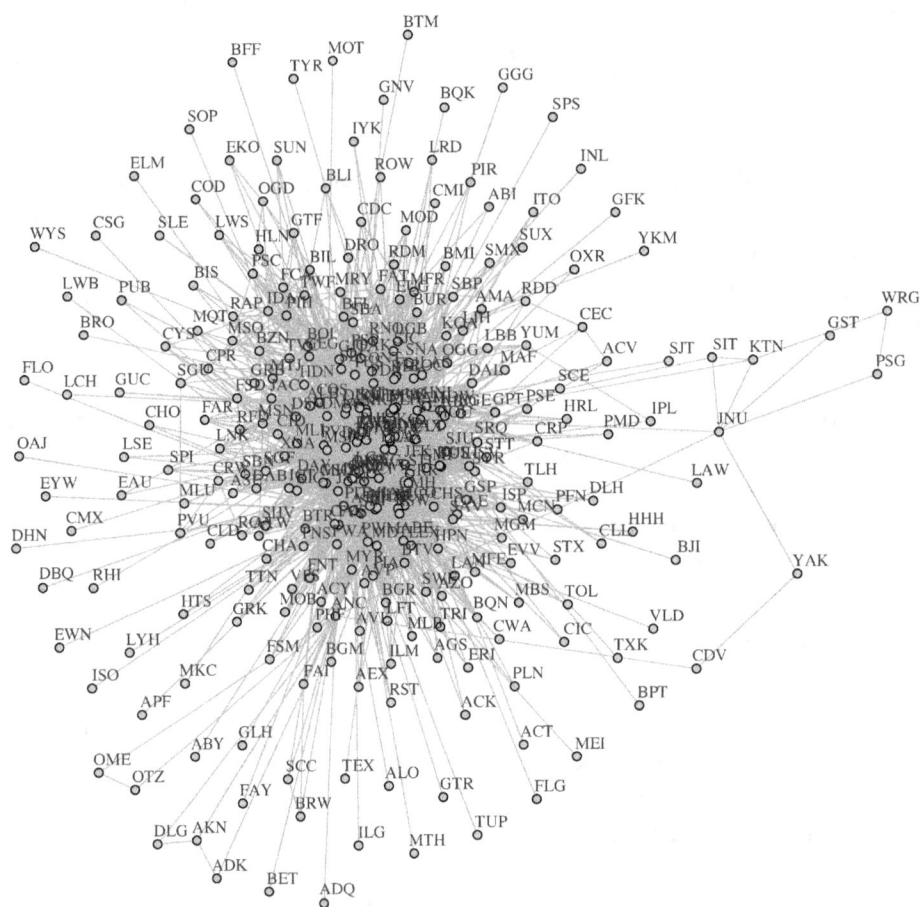

Figure 3.28: Network visualization of the flights between all United States airports in 2007. The airports are the nodes and the flights are the arcs. The result is a classic "hairball" that does not communicate any useful information about the U.S. airport network. The figure was created using the `igraph` package in R.

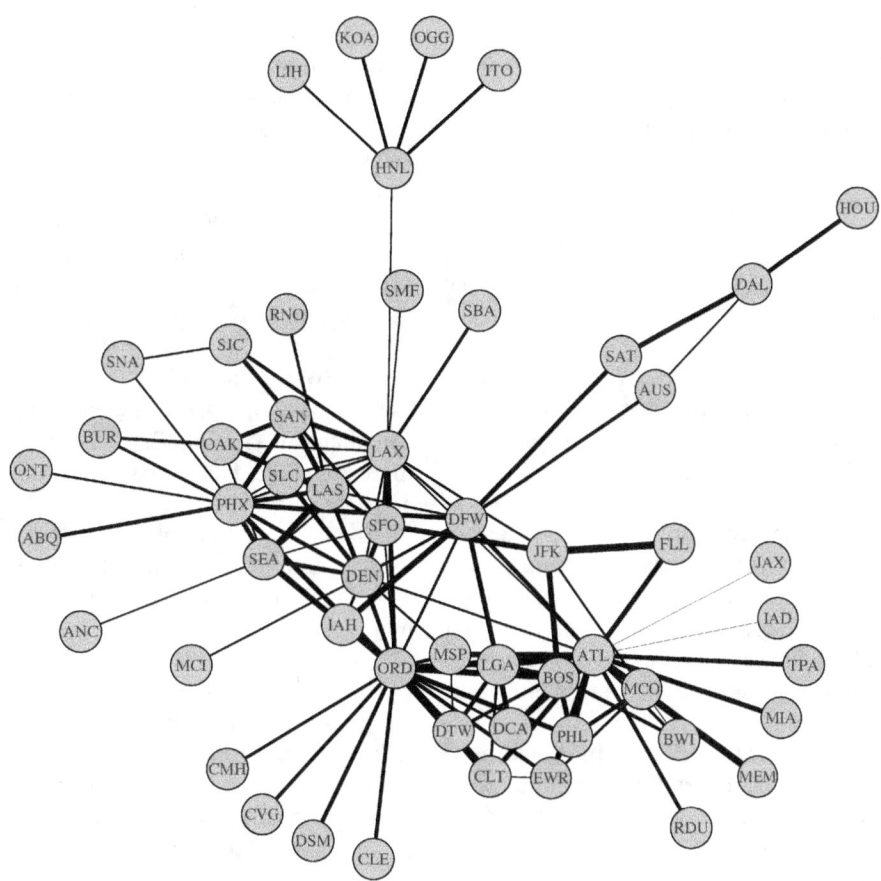

Figure 3.29: Network visualization of those United States airports with 10,000 or more flights in 2007 between them. The thickness of the arcs is proportional to the number of flights between the associated airports.

is necessary to figure out how to simplify the network display in order to learn about or communicate some important aspect of the network.

Network theory and the associated mathematical details are beyond the scope of this text. However, sometimes very simple and direct approaches can result in useful network visualizations. For example, Figure 3.29 is a display of the same 2007 flight data, using the igraph R package, but here only airports (nodes) with 10,000 or more flights originating or departing are plotted and the width of the arcs between airports is proportional to the number of flights along that arc. This display of the data now starts to show the interrelationships among the high volume airports and thus the core of the network.

The lesson here is that, when visualizing networks, it may be important to either do the visualization in interactive software so that the user can explore the network appropriately or to subset or highlight the data so that structure is visible. Figure 3.29 is an example of the latter approach in which the most connected nodes and associated arcs are highlighted.

3.5 Problems

Note: *For all plots and graphs in these problems, turn in a plot with informatively labeled axes and be sure to appropriately format, label, and title your plots. A good data visualization should communicate exactly what is being plotted and, to the greatest extent possible, the graph should be easy to understand and interpret.*

Problem 3.1 In your own words, describe how and why data visualization can be useful.

Problem 3.2 Find three examples of good data visualizations in the popular media or on the internet. Discuss what the visualizations show and explain why you consider them good visualizations.

Problem 3.3 Find three examples of poor data visualizations in the popular media or on the internet. Discuss what the visualizations are intended to show and explain why they are poor visualizations. Then, describe an alternate way to visualize of your examples and justify why your visualization is better.

Problem 3.4 In R, plot the data from Problem 2.5 as a histogram and a bar chart. Compare your plots, both to each other and to the summary statistics from Problems 2.5, 2.7, and 2.9. Which one do you prefer for visualizing this data and why?

Problem 3.5 In R, plot the data from Problem 2.5 using a boxplot. Identify the parts of the boxplot and label them with the appropriate numerical values taken from the summary statistics calculated in Problems 2.5, 2.7, and 2.9. (If you want to label your plot within R, the `arrows` and `text()` functions will be useful.) Comparing the boxplot to the histogram of Problem 3.4, which do you prefer for visualizing this data and why?

Problem 3.6 In R, plot the data from Problem 2.6 using a histogram and a box plot. On the boxplot, identify and label the parts of the plot with the appropriate numerical values taken from the summary statistics calculated in Problems 2.6, 2.8, and 2.10. On the histogram, identify where on the plot the same numerical values are located. (If you want to label your plot within R, the `arrows` and `text()` functions will be useful.) Compare your plots and explain which you prefer for this data and why.

Problem 3.7 In R, plot both the raw data from Problem 2.6 and the $\log_{10}$-transformed data using a box plot. On each boxplot, identify and label the parts of the plot with the appropriate numerical values. (If you want to label your plot within R, the `arrows` and `text()` functions will be useful.) Compare your plots and explain which you prefer and why.

Problem 3.8 In R, plot the data from Problem 2.11 using a scatterplot. Compare the plot to the correlation calculated in Problem 2.11. Do you learn anything more about the data from the plot?

Problem 3.9 In R, plot the data from Problem 2.12 using a scatterplot. Compare the plot to the correlation calculated in Problem 2.12. Do you learn anything more about the data from the plot?

Problem 3.10 In R, plot the data from Problem 2.13 using a scatterplot. Compare the plot to the correlation calculated in Problem 2.13. Do you learn anything more about the data from the plot?

Problem 3.11 Either manually or in R, appropriately plot the longitudinal data from Problem 2.14. (If you are using R, remember to use the `type="l"` option to create a line plot.) Now, overlay the moving average calculated in Problem 2.14 and the moving median calculated in Problem 2.15 (where the `lines()` function and the `lty` option will probably come in handy). What do you learn from the plot?

Problem 3.12 Either manually or in R, appropriately plot the longitudinal data from Problem 2.16. (If you are using R, remember to use the `type="l"` option to create a line plot.) Now, overlay the moving average calculated in Problem 2.16 and the moving median calculated in Problem 2.17 (where the `lines()` function and the `lty` option will probably come in handy). What do you learn from the plot?

Problem 3.13 Using R, plot a histogram and a boxplot of the 2007 aircraft flight arrival delay data (contained in the *ArrDelay* variable). As necessary, also plot the data appropriately transformed. In conjunction with the descriptive statistics calculated in Problem 2.18, discuss what you have learned from the plots and the descriptive statistics.

Problem 3.14 Using R, plot a histogram and a box-plot of the 2007 aircraft flight times (contained in the *AirTime* variable). As necessary, also plot the data appropriately transformed. In conjunction with the descriptive statistics calculated in Problem 2.19, discuss what you have learned from the plots and the descriptive statistics.

Problem 3.15 Using R, plot 2007 flight arrival delay (the *ArrDelay* variable) versus aircraft flight time (the *AirTime* variable) in a scatterplot. Compare the plot to the correlation calculated in Problem 2.20 and describe your results.

Problem 3.16 Using R, plot 2007 aircraft flight time (the *AirTime* variable) versus the distance flown (the *Distance* variable) in a scatterplot. Compare the plot to the correlation calculated in Problem 2.21 and describe your results.

Problem 3.17 Using R, graphically display the counts from the table created in Problem 2.22 appropriately in a bar chart.

Problem 3.18 Using R, graphically display the percentages from the table created in Problem 2.23 appropriately in a bar chart.

Problem 3.19 As described in Problem 2.24, the women data set comes with the base package in R. It contains the heights and weights of 15 women. (Type women to view the data and ?women to get some information about the data.) Graphically summarize the data using the appropriate plots and the discuss what the plots show.

Problem 3.20 As described in Problem 2.25, the faithful data set installed with the R base package has data on 272 eruptions of Old Faithful. The data set contains the length of each eruption (in minutes) and the time between eruptions, also in minutes. (Type faithful to view the data and ?faithful for more information about the data.) Graphically summarize the data using the appropriate plots and the discuss what the plots show.

Problem 3.21 As described in Problem 2.26, the cars data set installed with the R base package has data on 50 observations of car speed (in miles per hour) and distance to stop (in feet). Type cars to view the data and ?cars for more information about the data. Graph the data appropriately and describe what you learn from the plot.

Problem 3.22 As described in Problem 2.27, the mtcars data set installed with the R base package has data on 32 cars from the early 1970s. Type cars to view the data and ?cars for more information about the data. Graphically summarize each of the variables in the data appropriately using one or more plots. Using a scatterplot matrix, determine which of the other 10 variables is most positively correlated with miles per gallon (*mpg*) and which is most negatively correlated. (Hint: The plot() function run on the whole dataframe will produce a scatter plot matrix on all the variables in the dataframe – though note that with 10 variables each individual scatterplot will probably be too small to view.) Using the scatterplot matrix and other plots, what can you say about the data?